The Science Of God Volume 2

R Lindemann

Aleph Publications
Wisconsin, USA

The Science Of God Volume 2
Day Three - The Progression of Plants
Copyright 2016 - R Lindemann ©
All Rights Reserved. Published 2023

Aleph Publications
Manitowoc WI

All rights reserved. No part of this publication may be stored in a retrieval system, reproduced, or transmitted in any form, electronic, mechanical, photocopying, recording or other, without first obtaining the written permission of the copyright owners and the publisher.

Paperback Edition
ISBN13: 978-1-956814-26-2

33 32 31 30 29 28 27 26 25 24   2 3 4 5 6

Disclaimer

All information, views, thoughts, and opinions expressed herein are those of the author(s) and are being presented only for your consideration and should not be interpreted as advice to take any action. Any action you take with regard to implementing or not implementing the information, views, thoughts, and opinions contained within this published work is your own responsibility. Under no circumstances are distributor(s) and/or publisher(s) and/or author(s) of this work liable for any of your actions.

Anyone, especially those who have been victim of misdirected explanation and understanding, may be best served seeking wise counsel before deciding to implement any information, views, thoughts, opinions, or anything else that is offered for your consideration in this work. All information, views, thoughts, and opinions in this work are not advice, directive, recommendation, counsel, or any other indication for anyone to take any action. All information, views, thoughts, and opinions offered herein are offered only as suggestions for your personal consideration, which is done of your own free will. Your life is your own responsibility; use it wisely.

Any use of trade names or mention of commercial sources is for informational purposes only and does not imply endorsement or affiliation.

Please note that most of the items in quotes in this book are from various versions of the Bible and may have been paraphrased.

Dedication

This book is dedicated to anyone who wants to break free from wrong information to be able think freely while searching for what is true.

Contents

Chapter 1
Scientific Concepts and Distractions 1
    Scientific Complacency 4
    Referential Reasoning 5
    Scientific Oppression 7
    What We Thought in the Past 8
    Christian Perspective 10

Chapter 2
Existing In Geological Time 13
    Timeline Manipulation 13
    Rapidly Changing Times 16
    Imagine a Stack of Paper 18
    Interpreting the Bible 19
    If You Believe Lies You Have to Be Evasive 22

Chapter 3
Let the Earth Bring Forth 25
    A Dead Planet 29
    Plants Arise 32
    Rising Higher 33
    What Does the Bible Actually Say? 36

Chapter 4
Defining the Argument 39
    Capillary Activity 41
    Proving Our Hypothesis 44
    Using Bad Examples 44
    Finding Our Way 45

Chapter 5
A Need for Oxygen 47
    The Burden of Proof 47
    Insufficient Evidence 50
    The Evidence is Waiting to Be Found 51
    Proof versus Predictions 54

Chapter 6
Does Earth Plus Sun Equal Plants? 55
    The Order of Creation 57
    The Fertile Earth 61
    Filling a Purpose 62
    Working with what Plants Have 63

Chapter 7
The Protective Earth 67
    Within the Womb of Earth 68
    Enclosures of a Seed 71
    Abiding In Harmony 72
    They Found Their Home 73

Chapter 8
A Time before Sun 77
    Dividing Cells 78
    Multiplying Cells 80

    Us versus Them 81
    Separate the Issues 82
    Should Church and State Be Separate 85

Chapter 9
Protected Seeds 89
    The Bosom of Earth 89
    Embracing of Earth 90
    Every Plant for Itself 94
    Plants Within Plants 96

Chapter 10
Plant Life is Committed 99
    What Are We to Believe 99
    God's Handy Work 101
    A Commitment from God 103
    After Their Own Kind 104

Chapter 11
Open and Receptive 107
    Opening to See the Light 111
    Receiving Pollen 113
    Ears of Corn 114
    Why Are Most Plants Green? 115

Chapter 12
Calibration 119
    Astronomical Measurement 120
    Two Plus Two Will Never Equal Five 123
    Mile High Review 125
    Layers of Deceit 127
    Offensive Science 128
    Let Logic Be Your Guide 130

Chapter 13
So Many Plants 133
    The Flow of Life 134
    A Sea of Grass 135
    Who Will Care for the Plants 136
    Where Did the Water Come From? 138

Chapter 14
A Plant's Life 143
    Plant Activity 143
    Living Plants 145
    Plant Offspring 145
    Death of a Plant 146

Chapter 15
Self-Replication 147
    Funding the Research 147
    And the Like 148
    The Firmament 149
    The Evidence Supports It 150

Chapter 16
Perspective of Plant Evolution 153

Out of the Lips of Scientists	154
Claims of Statistically Verifiable Evidence	154
Choosing Your Evidence	155
Dolostone	157
The Black Layer	158
The Earth is a Living Document	159
Your Scientific "Ah-Ha" Moment	160

Chapter 17
Spreading the Seeds 163
Being First to Share	165
Prohibition of Discussion	166
Do Scientific Journals Tell All?	166
Explaining Yourself to Get Published	168

Chapter 18
A Plant's Harvest 171
Desire to Know with a Desire to Grow	172
The Joy of Plants	172
Gathering the Evidence	173
Reaching for the Stars	174

Chapter 19
Retaining the Information 177
Storing Up Information and Building the Instructions	178
Thinking Outside the Box to Think Inside the Box	178
Remembering the Evidence	179
Following the Instructions	180

Chapter 20
The Sprout 183
The First of Its Kind	183
Plant Messages	184
Plant Husbandry	185

Chapter 21
What Came First? 187
They Came First	188
Their Purpose	189
What Plant's Mean to Other Life	190
What Plants Mean to You	191

Chapter 22
We Must Ask Why 193
Did You Ever Wonder?	193
Seeking Answers	194
Asking the Right Questions Versus Asking the Questions Right	195
Seeking the Light	196

Chapter 23
Our Choices and Plant Choices 199
Anything for a Buck	200
Anything for God	200
Anything for the Agenda	201
What is Provable?	202

Chapter 24

Plants Live On **205**
   How Do They Even Bother? 206
   Why Do They Even Bother? 206
   Continuing Life 207
   The Plant Promise 208

Acknowledgements

Thank you to all those who have assisted in editing and with whom I had discussions about the topics covered in this book. But most importantly, I would like to acknowledge the abundant contributions to science and to this book by people who have long since passed. These courageous souls, in effort to serve truth, had sacrificed their reputation only to die with ridicule and without recognition for their vast contributions to science, and then only be recognized and properly acknowledged many years later long after they passed. Thank you all!

Introduction

The Creation story has become a mish-mush of varying beliefs ranging from it all being a myth, to the Bible being an inerrant accurate spectrum of events ranging from six-day-Creation across the spectrum all the way to Biblical-big-bang together with Biblical-evolution. And then there is the non-Biblical, or godless, approach using modern theories of solely a big bang and evolution with no Biblical or God connection whatsoever. How do we approach these topics without alienating readers? How does someone get people on all sides of the discussion to stop for a moment and consider the perspectives that others have proposed, because that's really the elephant in the room.

To start, we have to all shut up and listen *before* we speak. It is very disheartening when listening to so many people offer "opinions" when their opinions are not at all studied, and they only repeat things other people have said without actually having considered the problems that are inherent in any one theory or belief that they repeat. One point being made here is the views that many of us have regarding the Creation story are nothing more than unsubstantiated beliefs no matter which typical creation version we have chosen to adopt. It's not that the general theory is in error, such as it big banged or it was Created by the Creator, but rather the details of either belief are often greatly in error in regard to reality and science.

There is a simple reality which is that either everything was intentionally Created by the Creator, or it was not. If half of us believe one perspective to be true and the other half of us believe the other perspective to be true, then potentially, half of us are right and half of us are wrong. Determining who is right and who is wrong can only be done when we are willing to review all of the evidence in an honest manner.

In its simplest form, the question is: Was it Intentional Creation, or not? However, the details are a whole additional topic. We have faith in our beliefs whether it's some scientific big bang type belief or intentional Creation, and we choose to believe our chosen view with little evidence, because that is what we are taught to believe or have irrationally *chosen* to believe.

Contemplating Creation with a rational approach requires that a few basic points be met. The first point is that you cannot disregard the Bible just because you have chosen to believe it is not accurate. The second point is that the details of Creation are critical for understanding as to whether Creation was intentional or if it was some random banged occurrence. Since either may never have occurred, it is the details we must consider and it is those details that can light the way for us as we look into this subject. If you close your mind to any point of view, then you are certain to believe lies and wrong information because it pleases you to do so. Always leave your mind open to consider other perspectives.

It's common for young people in school to be lured into the belief that if they reject the Bible altogether that then they are being "open-minded", which is a misapplication of the idea of being "open-minded". Often, students go to school and are told that they need to think freely, which indeed is true, but that does not mean that you should believe whatever the teacher says. All things must be tested and tried to see if they hold up against scrutiny—especially what teachers say. A true and good teacher will invite scrutiny of their ideas, and if they don't, then they probably are just repeating things other people said without fully understanding those things. It's sad to say that most people's beliefs do not, and cannot, hold up against real scrutiny, which brings us right back to the safe positions of either insisting it was intentional Creation, or insisting it was not, thus leaving our correctness up to the flip of a coin.

We can do better than the flip of a coin when we read the history that sits before each and every one of us every day we are alive. You can look up, and you can look down, and you will see vast amounts of evidence from which you can make a choice as to what seems logical to you personally. But you could still be wrong even with that basic assessment. Too often we don't want to have to think because it actually takes effort to think these things through to their logical conclusions. Instead, we choose to just believe, rather than identifying the evidence and defining it and *then* making some rough predictions and finally setting out to prove our overall assessment.

In other words, put in the effort and take the time to consider the details *before* jumping to conclusions that will cause you to believe lies and embarrassing wrong information that you will have to correct later on when you finally realize your errors. To put it bluntly, you are almost certainly wrong if you have not processed the basic information that even children can grasp. Consider the theories of big bang, six-twenty-four-hour-day-Creation, and evolution as their inaccuracies are exposed while you read on. Let's see if we can all rebuild our sand foundation using solid rock in its place.

Chapter 1

Scientific Concepts and Distractions

Did plants evolve the way that it is often taught in public schools, where a "Creator" is not required? Or were the plants deliberately "Created" as is often taught in parochial schools, churches, and by preachers?

Before we get into the finer details of the Third Day of Creation, let's prime our minds a bit in this chapter and the next by asking some questions and by addressing some issues of our humanness. First, please note that the term "godless-evolution" is often used in this book and is in no way intended to be a derogatory term, and it is required in this entire topic to separate a godless-evolution perspective from a God-guided perspective, which is a very important distinction to be made.

To begin to answer the question of the manner of arrival of plant life, we have to open our minds and peel back some layers of incorrect thinking, along with some layers of Earth. Understanding the origin and arrival of plants requires that we first define a line between conventional scientific reasoning with an evolutionary process, versus conventional religious reasoning utilizing a deliberate creative process. But is there even a line to be drawn here? And if there is, then is that line a definitive and clear line, or is it a fuzzy gray area?

Consider some key questions: The first question is, when did plants come to be? This question is not asking about a specific timeframe of thousands or million or billions of years, but rather asks the order of events in which the plants where brought forth in the whole of evolution or creation.

Second, how long might have it taken for plants to actually form? Did an apple tree start as some sort of single-celled organism and slowly over millions of years grow directly into an apple tree, or did it evolve from some other fruit completely different from an apple? Or, is it possible that the Creator made it spontaneously grow as an apple tree in a matter of hours?

We also have to ask if vegetation, such as fruit trees, like apple trees and orange trees, came from the same organism or root? Or did they each have their own starting organism? Or did the Creator maybe spontaneously create each kind as an individual fully-grown tree? Or is there maybe some other more rational explanation?

Regardless of plants occurring through evolution, versus occurring through directed creation, did only one starting tree or cell exist and then from that single unit all other trees of like kind descended and populated the entire Earth?

And the last of the most fundamental question is, do plants require the sunlight to exist? This is perhaps one of the most critical questions in the plant-origins discussion when trying to reason things out between science and the Bible's stated order of events.

If you had the opportunity to read *The Science Of God Volume I - The First Four Days,* you will recall that while the book covered the first *four* days of Biblical creation, it only glossed over the third day when the plants were brought forth because *Volume I* is directed at astrophysics. I skipped the finer details of the plants when writing that book because the formation of plants is a topic in itself. The order of events is perhaps the most important aspect of creation regardless of it occurring through evolution, versus a deliberately creative process. This is because some things cannot be without other things occurring first.

If we want to try to arrive at any intelligible answers for this topic along with the questions surrounding it, we must first be able to conceive in our minds, the separation of ideas and the possible blending of ideas. Anyone who has read *The Science Of God Volume I – The First Four Days* knows that the blending the big bang and creation is not realistic, and quite honestly doing so is somewhat ungodly due to the lies and/or inaccuracies that we must embrace to rationalize big bang ideology. In saying that, I am not referring to big bang versus Creation, but rather big bang ideology whether scientific or otherwise is utterly irrational due to its inherent inaccuracies and flouting of the laws of physics.

The answers to the basic questions surrounding the arrival of plants are very much dependent upon the order-of-Creation or order-of-evolution and what occurred before it, and whether that was a random happenstance or an intentional creative act. Or is there possibly some in-between that we might be able to consider?

We should all strive for accuracy in our assessment of the Creation-versus-evolution debate, because without accuracy we are wrong, in error, or are outright liars whenever we share our inaccurate information with others. Maybe we should consider making some sort of pledge to ourselves and to the world; Where if you are a "believer" you should consider a short prayer, such as: Great and wonderful Creator, offer me the accurate understanding, vision, and words to be able to share your glorious Creation with unbelievers, nonbelievers, and believers alike.

Or logically, if you don't believe in God then it really doesn't matter as long as we all get our recognition whether right or wrong.

The base of these questions is this: Does any of this matter to you or are you free to believe whatever you want without repercussions? There are so many distractions in the analysis of the Creation-versus-evolution topic that we often get sidetracked and distracted by red herrings, rather than sticking to each point of contention in the Creation-versus-evolution topic.

Scientific Complacency

A brief review of our humanity as it pertains to science. Complacency is a very dangerous part of humanity. We have seen the devastation caused by complacency of the populous during many of the past decades. Here in this book we are referring to scientific complacency, but just consider what happens with political complacency, such as with communism for instance. When an oppressive government rules and the people who are being dominated reach their breaking point, they eventually rise up and overthrow the new power that is typically far more oppressive than the previous government was. This generation will oppose the absurd thought that everyone should be equal regardless of their efforts. They will understand that a system based upon rewards for work-done is the best and safest and most productive way to run any culture, including science culture.

But sometimes this complacency can last for several generations, and then when subsequent generations come along they learn to expect that things will be a certain way, and that way is then "normal" to them because they know no other way. With communism, eventually people won't do their fair share of the work but they expect that they should be taken care of regardless of their level of efforts and contributions to society. Additionally, the care given to them is left wanting.

A generation that has to somewhat forge their own way in life has a value which cannot be adequately expressed. But once we live an entitled-life that is full of entitled-expectation, even if that expectation is inadequate, we then typically fail to be driven by the rewards that come with work, which is why communism fails. Communism is a failure because there is no reason to progress because the people are not allowed self-improvement because all people must be "equal" and treated equal regardless of their level of effort, except of course for those in positions of power.

This same thing is true in science when we wrongly believe that what is stated is what is. We wrongly believe that we can, in an entitled way, get by on what our predecessors discovered, but this is wrong and dangerous! This is much like with the Earth-centric pre-Copernican views of the Earth-centric heavens that has now in modern times infiltrated science through big bang theology.

Referential Reasoning

We cannot say "Since there is Creation then there has to be a Creator" because if scientists who are unbelieving can offer plausible origins of all things, then a Creator would not be required in their origin theory. However, if there is no conclusive proof to their origin theories, then those theories are nothing more than "theories".

"Referential reasoning" is not a bad thing, but it can be a very misleading thing when the initial points of reference are inaccurate, lacking, or are flat-out wrong. For instance, when considering the beloved "geological column" you will find discrepancies from scientific discipline to scientific discipline. If you were to chronologically order all of the conventional "wisdom" of pop-science and its proposed years of the various layers in the geological column and then compare that with all radiometrically dated items found in science, you will find that there are too many anomalies to keep track of.

Does this problem make the geological column wrong? Yes and no. When sediment is laid down, our human logic forces us to understand that the lower layers are laid down first and then each subsequent layer is laid down on top of each previous layer. There are few people who would be foolish enough to debate this fundamental logic. However, there can be unique isolated instances where major geological events may have overturned sizeable sections of land causing the layers to be reversed, but this would be an outlier rather than a common occurrence and the sizable sections are still quite small relative to any one continent or region—Think in terms of feet or yards rather than miles.

We also have to consider that the geological column could be missing many layers. The Grand Canyon is a good illustration of this. Since the canyon has obviously eroded a great deal, if then at some point in the future a cataclysmic event occurs that would bury this canyon and a future people where to then core down into the canyon through the more recently laid material and then speculate on their findings, they might not realize that thousands, or possibly millions or even billions, of years of history are wiped clean, and thus they might make some very inaccurate assumptions about the actual geological history of that area.

But let us not allow this potential error to inhibit our quest for understanding our environment. Rather, use this example to realize that we simply do not know, and largely cannot prove, some of the specifics that we claim to be "scientific fact". We can really only speculate using our best understanding. The message in this is that we simply do not know for sure and we must use caution when we make our speculations, especially when we follow other people's speculations. And, when we make those speculations, it should not be done in an oppressive manner.

Scientific Oppression

In the same way that the people of the American colonies declared independence and said no longer will we permit any financial oppression, so too should we all do the same and reject scientific oppression. Many of the people of the American colonies died fighting for freedom because their oppressors demanded they follow the King's unreasonable rules and taxation. Their newly won independence has been a great gift to the world. Oppression still exists today, but now we face new oppressors who claim to be "scientists" and who will set out to unfairly defame us if we dare to disagree with them.

We see this in our high-tech social culture, but it is all too often true in the scientific world as well, and it has now bled over into the political world that has hijacked science for their nefarious agenda. Oppressors will gather their minions to shout your ideas down and they will attempt to inhibit your speech and your theories in effort to stop you from sharing those theories. This is true in life and in science and has been occurring for centuries, but it is far more vicious today than in the past. And because there is no longer any vetting on speech in our contemporary high-tech world, scientific oppression and all other forms of oppression are much more pronounced today than in centuries past.

We must endure this oppression and work through it if we ever hope to step through it all in effort to move humanity on to the next level. The oppression will not stop until Truth wins the hearts of all people. When humanity fails to stand for the light of truth in all aspects of life, then oppressors will rule and subdue others so that they do not lose their false high positions. As an example, if you disagree with something as simple as the time-scale in the modern interpretation of the "geological column", you might very well be the recipient of some vicious attacks from pop-science icons and their minions.

Truth cannot be thwarted, but it can be obscured by the lies we each choose to believe.

What We Thought in the Past

In our modern arrogance, too many of us today believe that the people of ancient days did not know things, or only progressively learned them over many thousands of years. To a small extent this is true, but when they looked at the heavens they were really quite clever as you will find if you care to look into their work. It is only recently that we have really been able to advance to deeper levels of understanding, due mostly to our new-found ability to launch ourselves into space. We wrongly believe that things have only been known for the last two or three hundred years, while all evidence shows us otherwise.

Look at flat-earth theory for instance. We believe that people believed that the Earth was flat, yet ancient artifacts and actual records indicate otherwise. If in centuries past someone made an argument that the Earth was flat and the logic seemed indisputable, then it is likely that many people would fall for the errors and believe them to be true and subsequently promote flat-earth as true, but this does not make it true. Later, people looked at the actual evidence rather than someone else's theory and they found that the earth was not flat and has no edges to fall off of. At that point we believe that we figured it out and that past civilizations did not know what we know today.

But if we look beyond conventional wisdom, we find that this is ridiculous and arrogant. Evidence of past civilizations indicates that they understood that the Earth was round at minimum, and we can even deduce that many knew that it was spherical. Somewhere along the way, we disregarded the early information indicating a circular and spherical Earth, and so some people chose to believe that the Earth was flat due to their irrational use of logic. So, in our modern arrogance, we wrongly believe that we have only just figured it out in the past several hundred years and that all prior people believed in flat Earth. But Columbus knew that it was obviously not flat.

It is when someone has a theory that only *appears* plausible to any contemporary civilization that we humans deceive ourselves. But not all of us fall for these lies and/or inaccuracies. It tends to be the most prominent figures in any society who are able to advance any theory. Even though we see our past mistakes, we do not want to believe that we believed wrongly in our relatively modern era. Evolution is one of those beliefs that has overstepped its bounds in our minds.

Evolution is often associated with animals rather than plants, but the same general arguments apply to the belief of evolution of both animals and plants. Evolution is a convincing argument that has lured many of us into its many errors. Does this mean evolution did not occur? No, but it also does not mean that it happened as proposed by many who hold evolution as their religion. There is a big difference between something being able to occur in theory, and something actually occurring in reality. And that is the error point of long-age-evolution. Because predictions have been made, and then proved "true" in the eyes of the predictor, it gives an illusion that evolution has occurred as stated, but this overlooks a great deal of other possibilities, and generally, it overlooks a great deal of scientific evidence and ignores basic logic.

In truth, "evolution" is very fast, and while "it is possible" that all things evolved from a single life-source cell, the reality is most likely otherwise. Changes within kinds are obvious, fast, and clear, but progression from one kind to another kind is grasping a bit and is ambiguous and not distinctly displayable and currently not scientifically provable.

The speculation that science has filled in the blanks with largely works only in the imaginations of those who have proposed the various evolutionary theories. When you look at the actual evidence, things are somewhat different from what we are told is true. For instance, we see all of the ancient depictions of large dinosaur looking creatures and creatures with skinned wings like pterodactyls, and yet archeologically, we ignore the information in these ancient depictions and interpret them as conveying "myth". We fail to realize that what we see in such ancient depictions is possibly what actually was witnessed by the eye of the people of that time who very well may have been recording what they saw. Additionally, the speculated timelines of when the drawings may have been done does not match pop-science's sacred geological column's timeline.

Consider our modern documentaries for instance. If a far-in-the-future culture sees one of the documentaries, will they interpret it as some sort of outlier because the subject of the documentary is no longer in their culture, or might they understand that it was telling them something that we actually experienced? If people of the future found bits of our civilization through our movies, they might think that we were murderous and that death loomed everywhere. But neither perspective makes the future people's assessment of us fully accurate. What we have chosen to believe about the past is most certainly erred on many points and these erred points can lead to us making many other wrong conclusions about the past.

The idea of evolution has been turned around and is presented backwards. To believe that all things evolved without design is perhaps the most ignorant stance mankind has ever witnessed. Does this mean that things cannot change through evolution? No, quite the contrary, in fact, whenever two organisms interbreed the result is always a cross between the two entities. This is very fast and occurs in one single generation. The so-called "survival of the fittest" is also likely an intended element of design.

Christian Perspective

Science's view of a conventional Christian perspective is that God did it all and created all things in six twenty-four-hour days, but there are quite a few versions of the perspectives on deliberate-creation. Some people do actually believe that God made **all** plants just as we see them today and did so in a single twenty-four-hour day, but we know that this is not true because **we** have crossbred flowers to make other varieties that likely never existed before. At the opposite end of the Creation spectrum there are some people who believe that God did everything using a big bang and evolution, but here again we can be reasonably certain that this is also not completely accurate when considering the lies and errors embodied in the various big bang theories alone.

A true "Christian" or true believer perspective must use science *and* the physical history, or it will end in failure as was all too often witnessed in the past. But using "science" to make our determinations does not mean that conventional pop-science beliefs are accurate or true, or that those scientific beliefs must be accepted in order to explain things through science using a Biblical perspective. We must consider the views, thoughts, and opinions of others, but it is each our own personal responsibility as to whether we will believe those views, thoughts, and opinions. No one can do it for us and we have to live with the consequences of the choices of what we each choose to believe and subsequently follow. Nowhere is this more prominent than in the sciences and religion.

There are ways to interpret scientific information that do not discredit the Bible. If you are a scientific Bible researcher your task is much more difficult in the eyes of pop-science, because you must follow *their* interpretation rules, but this is not true. Pop-science's interpretation rules are very inaccurate to a point where they have perverted the Genesis One text to a point of utter distortion and ruin.

When you accurately interpret Genesis, it is easier to find the truth because the fundamental framework is already done for us and we only need to fill in the blanks with the geological record that we find. And in doing so, we do not need to ignore or alter any actual scientific finds; however, we do have to disregard most of the pop-science conclusions.

Chapter 2

Existing In Geological Time

If we ignore the two rare examples given in the last chapter, one of overturned land and another of a canyon being filled in, the sediment layers in the "geological column" can be quite telling of the order of events that may have occurred. In using Earth's visible strata in places such as the Grand Canyon and various core samples, we can in most cases be reasonably certain that the lower levels on the geological column came before others, meaning that they occurred first, and the closer to the top of the column the layer is then the newer the layer is. So we have this existing record of the order of events, but in what timeframe did those events occur? And what events are specifically recorded in that solidified data?

Timeline Manipulation

If we say that the geological column reflects only several thousands of years of information, then the long-age big bang crowd gets upset. Or, if we say that the geological column reflects millions or even billions of years, then the six-day-twenty-four-hour creationists get upset. But are either of these two interpretations of the same data proper and true?

There is an enormous disconnect between some religious creation beliefs and science that any truly honest creation believer knows cannot exist in reality. But there is also a disconnect between "science" and science itself that is often ignored, and too many people of "science" refuse to acknowledge that any disconnect exists. The most notable stumbling block in science is the gross misinterpretation of data by the various scientific disciplines regarding the geological column and even the determining of species.

There is a simple truth that no amount of demanding or proselytizing can thwart, and it is the simple fact that all timelines *must* agree with relative accuracy through all scientific and biblical disciplines. If they do not agree then something is wrong with the interpretation of some or all of the data.

What we want to believe is irrelevant as to what actually occurred, and, for myself, I prefer to be accurate to reality rather than believing something because it feels good. *Truth* is far more important than what any of us *want to believe* about anything.

Timeline manipulation becomes a problem because some proponents of the Bible fail to see that opponents of the Bible manipulate timelines and disconnect them entirely in order to further their own theories and scientific agenda. Some people do this deliberately; however, most people manipulate timelines because they simply do not understand that it *must* agree through *all* scientific disciplines. But some Bible proponents will use potentially erred dating to "prove" the Bible to be true to the interpretation that *they* have chosen to agree with based upon their chosen erred scientific geological dating index.

All timelines share one common trait, and that trait is reality. If it actually occurred, then it is in the timeline somewhere, but if it did not occur then it does not have a timeline and any suggested timelines that include the imaginative events are imaginative and ultimately false regarding those particular imagined events. We need to initially disregard any counting of years and first look at the actual events that would most likely have caused the formation of the various layers, and the plausibility of any theories as to how those layers formed.

We witness timeline manipulation in regard to historical events even though those events have been recorded in written accounts. Yet, a "scholar" will dig up some artifact and proceed to have it dated and then their newly proposed age of the archeological find becomes "indisputable fact" to them, even though the time ranges given can be plus or minus upwards of ten-thousand years. With some scholars, there appears to be a distinct effort to discredit any references to the Bible that might invalidate their newly invented timeline, and so, the events of the Bible do not reconcile with their invented dates. Thus, proving in their own minds, and unfortunately in the minds of many others, that the Bible does not fit into *any* timeline.

I come from a world of measuring in ten-thousandths of an inch where if you are inaccurate with your measurements then things generally will not work as intended. However, in the world of astrophysics, geology, and archeology, accuracy is of no matter. Often provably erroneous numbers for time are thrown out to an unsuspecting public, with many of these erroneous numbers having inaccuracies upward of plus or minus twenty percent. There are few industries that allow that sort of mathematical tolerance. We may need to redefine the terms "scholar" and "scientist" at some point in the future to compensate for such inaccuracies.

When attempting to understand the historical events of Earth's creation, regardless of how it occurred, we must realize that everything either fits fully and completely on a shared timeline, or it simply did not occur. There are no alternate realities, and there are no separate timelines. We all share the same timeline and we also share the same timeline that all of history shares. We are just at a different position on that timeline.

When opponents of the Bible speak of history, they often regard the Greeks as the foremost thinkers, as if the stories and writings of the Greeks are all real and actually existed and are not myth, while at the same time they will disregard the Bible as mere fairytales. This is either very disingenuous or very ignorant thinking considering the extensive backup documentation that is found world round regarding Biblical texts that far exceeds any other ancient culture and any other book.

But regarding timelines, in this book we are not concerned with human history at all because the account of the creation of plants came before people even existed, according the Genesis Creation account. It's clear that we humans have flaws in our assessment of historical timeframes, so we're going to return our focus to the first four days of Creation. And more specifically, to the third day that is responsible for the advent of vegetation which includes all plant life.

Rapidly Changing Times

We all know that Earth changes over time, but it is our assessment of the duration of time that various changes occurred that causes division in the different scientific disciplines, including Biblical science. Interestingly, the science-world tends to indicate that the various layers of sediment have been laid down over thousands and even millions of years for each sediment layer, but this is highly unlikely. Any layer of sediment laid down for more than a couple of years would show many signs that it had been there for a relatively long period of time, and more notably, it would be highly unlikely that fossils would be buried within the sedimentary layer unless that layer was laid down in a matter of days or weeks.

Anyone who has ever walked through the woods and witnessed a dead animal can attest to the fact that the animal-remains tend to get eaten and scattered quickly. And if you have had the opportunity, you will find that the bones will relatively quickly turn to dust or be scattered in only a few years' time when exposed to the elements and other creatures. This is clear indication that the majority of the fossils laid down had been buried in very short order versus the pop-science estimates of the time it took to lay down each layer in thousands, tens of thousands, even millions of years.

But a more damning indicator of the current scientific calibration of the geology comes in the piercing of sedimentary layers from which many larger fossils are extracted. As we regularly witness today in only a few years' time regarding the death, decay, and dispersion of an animal carcass, we know that it is an utterly absurd notion that a whale or dinosaur carcass would have been able to withstand the time scale that would have been required to be exposed through even a single layer, let alone several layers, of sediment according to the conventional per-layer timescale from pop-science's geological estimations.

Simple logic alone should tell you that there are problems with the pop-science calibration of the geological timeline. I only mention animals here because they are large enough to lend witness to the misinterpretation of the pop-science timeline. But, we also have tree fossils that pierce even more strata than do any animals. This often-ignored information is perhaps the single most damning bit of evidence of the conventional calibration of the pop-science geological timeline. If you happened to have grown up in rural areas of the countryside and have lived long enough, you will have witnessed how quickly smaller topography can change from natural causes without any cataclysmic activity.

Small plants typically do not offer the same bit of insight into the actual rapidity of geological formation. This is because small plants, and more notably their leaves or pedals, are often trapped between sediment layers. Because they do not pierce through more than one layer and are often caught between two layers, they can only speak to the time-frame order in which they have been encapsulated. But even though human logic should tell us that each layer likely occurred in short order, we still do not know the actual timescale of the deposition of the layer or the timespan *between* the depositing of each layer. While each layer is likely not billions, millions, or even thousands of years of sediment, there is still a great deal of speculation in attempting to establish a reasonably logical timescale for the actual geological column.

Imagine a Stack of Paper

An example that I like to use from my previous books to help to illustrate time and our place within that time is a very tall stack of paper representing the age of the Universe. Let's imagine a single stack of paper stack standing upright with one piece of paper lying flat upon the next. Each piece being only about .004 of an inch thick (4/1000), which is about the thickness of a typical piece of paper you might put in your computer printer on your desk, or a piece of typical typewriter paper if you are familiar with one of those devices. When doing the math, you will find that at this thickness of paper one inch of paper on the stack will be 250 sheets. A one-foot thick stack of this paper will be 3000 sheets. And stack one mile high will have 15,840,000 sheets. So as to not miss any zeros, that's nearly sixteen million sheets in the stack. A billion sheets mathematically calculate to be a stack just over 63 miles tall.

So as to keep this within an easy-to-grasp visual we will imagine a stack of paper only one mile tall containing only 15,840,000 sheets of paper. If the supposed age of the universe is thirteen to sixteen billion years old, then for illustration purposes here, we will place it at fifteen-billion Eight-hundred-forty million years, thus making each paper in our stack represent one thousand years.

With the average human life being about 77 years, (we'll call it one hundred years for ease of use) that makes a human lifespan equal to one tenth the thickness of a single piece of paper, or four one-ten-thousandths of an inch (4/10,000 or 0.0004), which is about the thickness of typical kitchen plastic-wrap film you might use to wrap up your sandwich. That thickness is *your entire lifespan* in the grand scale of a one-mile-high stack of paper in relation to proposed age of the Universe. Now, the Earth is claimed to be roughly a quarter of that stack in age so we can remove three quarters of the stack of paper so that the remaining stack is only 1320 feet tall, or a quarter of a mile. But *your entire life* is still only represented by your plastic wrap thickness which is about 1/10 the thickness of a sheet of paper. So, in relation to the proposed age of the Earth, the entirety of "modern science" is not even a single paper's thickness. It is just under half the thickness of a single paper. Now imagine three or four layers of plastic wrap representing about one-hundred years for each layer and imagine that in comparison to the quarter mile high stack of paper.

Our current scientific conclusions are based solely upon our collective experience of time over the past several hundred years. And if we imagine that our timescales that often proclaim a plus or minus twenty percent accuracy have much basis in reality, we are likely mistaken.

Interpreting the Bible

Biblical misinterpretation is a big part of the reason that scientific atheists have hijacked science from creationists, and rightfully so. The problem with Biblical followers is not that they are religious, the problem is that all too often they incorrectly interpret the text of the Bible. And worse, they often attempt to meld it with faulty speculative pop-science theories.

But many people of science are no better in their own interpretation of what the Bible says. This is very disappointing, because if anyone should be able to accurately understand Genesis One it should be the science world, yet they do not.

Many creationists only believe what they believe because that is what they were taught while growing up or in college, which is understandable if they are not actually researching the scientific aspects of Creation. But people using the hallowed "scientific method" have failed repeatedly to actually get it right in regard to the Bible's accounting of Creation when they insist that the Bible says it all happened in six twenty-four-hour days.

While it is a noble effort to attempt to reconcile science and the Bible, it is not so noble when you include science's most prolific errors. Many people just read the Creation text and say "Oh that's just the way it is" rather than asking why it is stated the way it is, as it is accounted in the Bible. Or they attempt to invent potential methods of creation that can, in their own minds, reconcile with pop-science.

But when someone actually takes the time to accurately and unbiasedly read authoritative versions of the Genesis One text, then their entire interpretation of events will change provided that they can shed their preconceived ideas. We often will not let go of our preconceived and preconditioned interpretation of the Bible in order to actually get the accurate picture. This does not mean that people should not use the Bible as a reference tool, but rather to more carefully do so. Science is guilty of the same thing as religion is, and in fact, science has become a religion in itself for most pop-science-minded people as well as many others from the general populous. Too often we get caught up in the argument itself and decide to take sides and stand firm, unwilling to receive other information. Our problem has become, not so much what is true or what is not true, but rather our problem is that we *demand* that our own ideas or interpretations are "right" as we attempt to see how it all fits together.

As discussed in *The Science Of God Volume I - The First Four Days*, the version of the Bible you choose to get your information from will have an enormous impact on what you are able to ascertain from the text. If you look at the history of the steps that occurred to arrive at our modern Bibles, you will quickly find that they all trace back to a few critical collections of documents, which are the Hebrew, Greek Septuagint, Aramaic, and Vulgate texts. The most prominent is the Latin Vulgate produced just before 400 AD which was an effort to standardize the texts and codify them so that interpretation deviations from the text that occurred before that time could be stopped. The Vulgate stood the test of time and was partially used along with any other available ancient texts in the 1500s and 1600s to translate the Bible into relatable languages of the people, namely English and German. If you want to know more about this you can read more in the book *Understanding The Bible - The Bible How-To Manual AND The Things We Don't See*.

Around the time of the Reformation revolt, King James commissioned that Bible to be translated in to English for what became the Church of England. Around the same time, the Douay Rheims English version of the Vulgate was also translated, as well as the German Luther Bible. This all occurred around the same time as the Gutenberg printing press was invented and the Jerome Latin Vulgate was mass printed and is known as the Gutenberg Bible.

Using the early printing press was a painstaking effort. However, it was far faster and could more consistently duplicate the Bible than transcription by hand alone could. The speed at which a Bible could then be produced on the new printing press allowed larger scale distribution of these versions of the Bible. The Hebrew texts and the Septuagint were the lions-share of the source material for Jerome's Latin Vulgate. The Vulgate stood as the Biblical standard for over a thousand years, until the creation of the Douay Rheims, The King James, and The Luther Bibles.

Does any of that matter? I suppose some people would claim that it doesn't, but when you consider the consistency between these versions, they become very impressively accurate. Are they perfect translations? Not likely, but when we get beyond our petty preferences of interpretation, we can take all of these older versions together and see what agrees and what does not agree between them. Then we can use basic human logic and scientific observation to determine what was likely intended in the text. And you must understand that there are relatively few discrepancies between these older versions. However, this changed in the 1800s when higher speed mechanized printing proliferated the ability to republish Bibles with new unauthorized verbiage.

While the efforts may have been for a noble cause, many of the translations done over the past couple of centuries have obscured many critical points in the Bible. When we take a version and replace a word such as "firmament" with a word like "sky" it causes a monumental shift in our perspective of the events described in Genesis One, thus hindering our ability to accurately interpret the text.

If You Believe Lies You Have to Be Evasive

When discussing pop-science, you will find that if someone has a belief, whether scientific or Biblical, and that belief has anomalies within it, then they will have to be evasive when discussing that aspect of their chosen belief. However, when you are truly open to new information and are willing to process that information, then you will readily invite new information in order to compare it to your own theories in effort to test them for accuracy. Whenever we build our conclusions on *false* premise, we get frustrated when being questioned about the conclusions of our theory built upon that false premise.

If you are being pressed in questioning about your thoughts and theories and the questioning is reasonable and not deliberately contentious, but you become agitated and frustrated and are unable to deliver sound answers to the person or people inquiring of you, then you probably should reconsider your theories and recheck them for accuracy. It is sometimes okay to simply state that you are unsure about certain aspects of your theory while you are studying and developing it. But if someone is presenting a theory containing obvious flaws and breaches as if the theory is "fact", then they are being dishonest.

Chapter 3

Let the Earth Bring Forth

Now that we understand that there may be some errors in our thinking regarding science and/or religion and our interpretation of evolution and creation, we can dive into the topic at hand which is the advent of plant life here on Earth.

As discussed in an earlier chapter, the version(s) of the Bible you choose to use for scientific analysis of the Genesis text, will greatly affect your ability to accurately analyze that text. This is especially true regarding the first four days of the Creation account which also pertains to the plants and all general vegetation. If you have not read *The Science Of God Volume I – The First Four Days* consider doing so, because it meticulously combs through the text of the first four days of the Creation account of Genesis One, but only lightly touches on day three. Days One, Two, and parts of Three, and day Four are all more about astrophysics than they are about plant biology.

The Genesis account of creation is unique in all of the Bible because, as a technicality, it is either invented and utterly irrelevant, or it was given to man by the Creator. If it was invented, then nothing really matters. But if it was given by a Creator, then it is a very important document that does not embody the human influence that the accounts of humanity do throughout most of the rest of the Bible. And while it is affected by us in translation, it is uniquely different in that we humans were not here to witness the described events. Our only option is to try to understand it as told or given to us. But in our modern versions we get into trouble in the very first verse of Genesis One. Notice the differences in the various modern Bible versions-in the very first sentence of Genesis One:

Douay-Rheims Bible (Originally largely translated from Septuagint and Masoretic then later revised with KJ version)
    1) In the beginning God created heaven, and earth.

JPS Tanakh 1917 – (Largely translated from Masoretic)
Webster's Bible Translation
English Revised Version
King James Bible – (Largely translated from Masoretic)
American King James Version – (Largely translated from King James Bible)
King James 2000 Bible – (Largely translated from King James Bible)
    2) In the beginning God created the heaven and the earth.

New American Standard Bible
Holman Christian Standard Bible
NET Bible
New Living Translation
Jubilee Bible 2000
American Standard Version
Darby Bible Translation
World English Bible
New American Standard 1977
    3) In the beginning God created the heavens and the earth.

GOD'S WORD Translation
    4) In the beginning God created heaven and earth.

Young's Literal Translation
    5) In the beginning of God's preparing the heavens and the earth.

English Standard Version
    6) In the beginning, God created the heavens and the earth.

International Standard Version
    7) In the beginning, God created the universe.

Do these differences matter? Yes and no. If you are casually reading the text you will get the general message and understand that some Creator Created it all, so it might not have much relevance to casual readers. However, when you are trying to understand the science behind it all, then some of the small things matter a great deal in interpretation. If you refuse to consider the various versions, then you are likely to run into scientific problems in the text that is found in some modern versions. In the event that you decided to choose one of the less authoritative versions, you can expect a great deal of mental conflict within yourself and with science.

To stay focused on the plants or any vegetation which this volume of *The Science Of God* is about, we will mostly be discussing days three and four, only touching on days one and two when needed. Use the Douay Rheims version as a research starting point and then call upon other older versions after the questions and answers are formed to see how it all compares from version to version. Disregard any versions that have replaced the term "firmament" with "sky, dome, horizon" etc, as those terms greatly pervert the text and hinder your ability to accurately interpret the text.

Douay-Rheims Bible (Largely translated from Septuagint and Masoretic)
1) And God said, Let there be a firmament made amidst the waters: and let it divide the waters from the waters.

American Standard Version
JPS Tanakh 1917 – (Largely translated from Masoretic)
English Revised Version
King James Bible – (Largely translated from Masoretic)
King James 2000 Bible – (Largely translated from King James Bible)
Jubilee Bible 2000
Webster's Bible Translation
2) And God said, Let there be a firmament in the midst of the waters, and let it divide the waters from the waters.

American King James Version
3) And God said, Let there be a firmament in the middle of the waters, and let it divide the waters from the waters.

Darby Bible Translation

4) And God said, Let there be an expanse in the midst of the waters, and let it be a division between waters and waters.

English Standard Version
5) And God said, Let there be an expanse in the midst of the waters, and let it separate the waters from the waters.

GOD'S WORD® Translation
6) Then God said, Let there be a horizon in the middle of the water in order to separate the water.

Holman Christian Standard Bible
7) Then God said, Let there be an expanse between the waters, separating water from water.

International Standard Version
8) Then God said, Let there be a canopy between bodies of water, separating bodies of water from bodies of water!

NET Bible
9) God said, Let there be an expanse in the midst of the waters and let it separate water from water.

New American Standard 1977
10) Then God said, Let there be an expanse in the midst of the waters, and let it separate the waters from the waters.

New American Standard Bible
11) Then God said, Let there be an expanse in the midst of the waters, and let it separate the waters from the waters.

New International Version
12) And God said, Let there be a vault between the waters to separate water from water.

New Living Translation
13) Then God said, "Let there be a space between the waters, to separate the waters of the heavens from the waters of the earth.

World English Bible
14) God said, Let there be an expanse in the middle of the waters, and let it divide the waters from the waters.

Young's Literal Translation
15) And God saith, Let an expanse be in the midst of the waters, and let it be separating between waters and waters.

As you can see the differences being spoken of here can cause considerable differences in our understanding the text.

Here is the Douay Rheims version of days three and four:

[9] God also said: Let the waters that are under the heaven, be gathered together into one place: and let the dry land appear. And it was so done. [10] And God called the dry land, Earth; and the gathering together of the waters, he called Seas. And God saw that it was good.

[11] And he said: Let the earth bring forth the green herb, and such as may seed, and the fruit tree yielding fruit after its kind, which may have seed in itself upon the earth. And it was so done. [12] And the earth brought forth the green herb, and such as yieldeth seed according to its kind, and the tree that beareth fruit, having seed each one according to its kind. And God saw that it was good. [13] And the evening and the morning were the third day. [14] And God said: Let there be lights made in the firmament of heaven, to divide the day and the night, and let them be for signs, and for seasons, and for days and years: [15] To shine in the firmament of heaven, and to give light upon the earth. And it was so done.

[16] And God made two great lights: a greater light to rule the day; and a lesser light to rule the night: and the stars. [17] And he set them in the firmament of heaven to shine upon the earth. [18] And to rule the day and the night, and to divide the light and the darkness. And God saw that it was good. [19] And the evening and morning were the fourth day.

In our modern culture we have become too accustomed to reading the Genesis text with its modernized punctuation, capitalizations, and verse numbers. I like to strip all of those out leaving only the pure text to read. So from this point forward, most of the Bible quotes will have the punctuation, capitals, and verse numbers removed, but nothing else will be changed.

A Dead Planet

As you read make sure to make the distinction between **planets** that move about the heavens and **plants** that grow from the earth because both are spoken of. We often view Earth as if it was a dead planet that slowly began to have bacterial activity and slowly formed nutrient enriched soil and then eventually plants began to form. But this is likely not true. Planets more likely become "dead" *after* life fails to take root on them because they do not reside within suitable distance from their star in combination with their spin-rate, mass, and planet composition. But let's stop right here and try to determine what a "planet" actually is.

A planet is a large celestial body that orbits a star. The term planet means wanderer or basically a moving star. Stars that emit light appear stationary to us but planets which only reflect light wander in the sky from night to night as they circle their sun/star.

So by that definition, the moons that orbit their planets could also be defined as "planets". What is the authority to name things? Who gets to do this? A hidden inclination in our hearts is that if we know the name of something, then we know about it. This is because knowing a name allows us to discuss it with others by merely mentioning the name of the thing or theory.

Who named the "atom" and why is it called that, and does *atom* mean anything? Should the name "Atom" be changed? Maybe. If a different name that is a better fit would allow people's minds to open up and better understand what we call an "atom", then renaming would have value. To most of us, scientific naming is a bit arbitrary and means very little to us, especially if it is named after a person. But many things are named appropriately and do describe the essence of the item or aspects of it in the language that was used to name the item. Just try looking up a term such as "atom" and you might find it to be very interesting and possibly a poor choice.

Planets being "wandering stars" is somewhat appropriate, but moon on the other hand is a bit different. The term *moon* and *month* are from the same root and it is likely that "moon" is the base of that root and that month or **moon**th (full moon to full moon) is the approximate time of a month for us here on Earth. However, "moon" in that case would differ a great deal from planet to planet if the planet actually has a "moon" because the rate at which those moons circle the planet differs from our own Earth's moon's orbit.

Moons are visible around other planets and they do wander, so technically they too could be considered wandering stars using the same logic as is used to derive the term "planet".

Getting back to the subject of dead planets, it is highly unlikely that planets began as the "dead" planets we see today such as Mars or our Moon. The likelihood is that they were somewhat active and lively but eventually became completely dead, such as we see with mars or our moon, or other planets and their moons that we have sent probes to. If a planet or any celestial body in space had enough gravity to balance the elements and compounds in such a way that life could advance, then it would advance. But if the celestial body was too small and too cold, or too hot, to sustain life, then life would have tried but failed. You might want to picture this as some sort of desert-like view with dead plants and animal skulls strewn about, but that is not at all how it would have been. Active plant life would never have developed that far.

In our contemporary times we have the wonder of modern engineering that allows us to launch ourselves and our devices into space to further investigate the heavens. And so far, no matter where we look, we see repeated patterns of past activity. Some planets and moons show that it is highly likely that they had liquid material flowing on them in large volume, causing washout from the liquid flow much like the Grand Canyon. Many also have had what appears to be obvious volcanic activity, some of which appear to still be active. We also see changing weather patterns on planets that have visible atmosphere.

But amongst all of the visible planets, whether planet, moon or asteroid, Earth is unique in our solar system and it is very much alive and was likely not ever a "dead planet". The thought that Mars could ever be a good place to live and could in anyway be better than Earth is really quite ridiculous. But this does not mean that we shouldn't investigate it or other planets, moons, and stars.

The **plants** or vegetation on our planet Earth, according to Genesis One were brought forth on the third day of Creation. As discussed in *The Science Of God Volume I – The First Four Days*, the chances of all of Creation being brought into existence in six twenty-four hour days is nonsensical because a "day" as we know it today, could not have occurred until at least the fourth day as described in Genesis One. So, the "days" spoke of in Genesis One must be understood as instances of events rather than twenty-four-hour days. While the *initiation* of those events might have occurred in a split second and thus could have occurred in a "day", the fullness of the event could have taken place over extended periods of time. This is true for all four of the first four days of Genesis One's Creation account.

Plants Arise

Just when and how did the plants come to be? Was there some primordial soup that was hit by lightning that created the first amino acids allowing the possibility for plant evolution to occur as proposed by science? Or maybe, did God just say "Let there be" and then POOF! the Earth was suddenly filled with trees?

Let's take a look at the pure text stripped of all capitalization and punctuation influences and see what it says about how the plants came to be through those first critical four days, and allow our logic to dictate the order of events rather than added commas and full stops:

"in the beginning god created heaven and earth and the earth was void and empty and darkness was upon the face of the deep and the spirit of god moved over the waters and god said be light made and light was made and god saw the light that it was good and he divided the light from the darkness and he called the light day and the darkness night and there was evening and morning one day and god said let there be a firmament made amidst the waters and let it divide the waters from the waters and god made a firmament and divided the waters that were under the firmament from those that were above the firmament and it was so and god called the firmament heaven and the evening and morning were the second day god also said let the waters that are under the heaven be gathered together into one place and let the dry land appear and it was so done and god called the dry land earth and the gathering together of the waters he called seas and god saw that it was good and he said let the earth bring forth the green herb and such as may seed and the fruit tree yielding fruit after its kind which may have seed in itself upon the earth and it was so done and the earth brought forth the green herb and such as yieldeth seed according to its kind and the tree that beareth fruit having seed each one according to its kind and god saw that it was good and the evening and the morning were the third day and god said let there be lights made in the firmament of heaven to divide the day and the night and let them be for signs and for seasons and for days and years to shine in the firmament of heaven and to give light upon the earth and it was so done and god made two great lights a greater light to rule the day and a lesser light to rule the night and the stars and he set them in the firmament of heaven to shine upon the earth and to rule the day and the night and to divide the light and the darkness and god saw that it was good and the evening and morning were the fourth day"

You will notice that if this text is written in the order of events it makes perfect sense. However, if the text is written in a retrospective manner referencing heaven or the sun before they existed, then it all seems bit foolish. And that is a critically important point to understand, especially regarding the creation of the plants.

Rising Higher

Very interestingly, the waters under heaven were gathered together into one place. And the dry land was then named "Earth" and the gathered waters were called "Seas". Just by this alone we are forced to picture gravity at work doing its thing to some extent. But the "Earth" at this point would likely have been very wetted with some sort of liquid and likely was warm and mushy, at least below the surface, thus creating a perfect environment for organic activity to begin. It is also very likely that considerable heat was generated during this event making the planet very warm, but since the "waters be gathered" we have to assume that the temperature was low enough that the water would not boil off into a steam state or that it was not cold enough that it would freeze solid.

Given the description in the text, we also have to understand that if the waters were gathered in such a way as to form the seas and dry land did appear, that there would have had to have been atmospheric pressure of some sort or all of the water would have boiled into steam due to there being a vacuum of space that would have surrounded the "Earth". However, it is very possible that gravity, as we think of it today, was not alone at work here and other things might have been at play even though gravity itself may very well have allowed the atmosphere to develop from the available elements at the time.

If the earth was a gigantic blob of mush and had not yet fully formed as we experience it today, then the gathered waters may possibly have not yet been in the H^2O form at the initiation of that event and thus would not have been susceptible to boiling in a vacuum environment. But as the various atoms were drawn together forming the various compounds, gravity dictates that an atmosphere would be the byproduct of that activity. Even if the water was in a cool vacuous steam state and drawn together by gravity, eventually the developing and increasing pressure would cause precipitation of that vapor allowing water as we know it today in the seas to have formed.

It is only at this point in the creation events that higher forms of creation could exist or begin to exist. From our experiences through recorded history, without water we can be reasonably certain that no plant could thrive or even begin to take root. Once the water had been separated out as H^2O in the "seas", it is at that unique point in the order of events that the plants were created by God... or were they? Were plants actually created by God or did it occur through some other means?

In general, the first two days are specifically deliberately creative events. However, the rest of the creation events of the first four days are more about the organization of the already created. During the first four days the only things that were specifically created are "heaven, earth, light", and "firmament", the remaining events of the first four days are more about making distinctions between things.

The making of heaven and earth in the first sentence is likely *not* the same Heaven and Earth that we think of when we discuss Heaven and Earth today. And the heaven and earth in the first sentence are possibly not a part of the first day, but may very well be a part of it because heaven, earth, and light are created items and if you read *The Science Of God Volume I – The First Four Days* you will realize that those three items, heaven, earth, and light are critical aspects that all things, seen and unseen, are produced from.

Then on the second day a "firmament was made in the midst of the "waters to separate the waters from the waters". But again, we have to understand what the "waters" are. Based upon the order of the text and the fact that "water" at that point was not a created item, we are left with the fact that "water" was likely a property of the created "heaven and earth" that can be best described as fluidity—it can move and flow.

On day three the waters under heaven were gathered, which is in essence a separation from what was not gathered. So, the gathered waters were not created as a specific act but rather were "gathered", and thus further separated from other things through that gathering. As you will see if you read the text, "god made a firmament and divided the waters that were under the firmament from those that were above the firmament" and then the next action is "the waters that are under the heaven be gathered together into one place". The waters below were further separated or organized through the gathering action. If we view the "waters" as a mass of fluid material consisting of many elements and then as those elements were drawn together to form compounds, the compounds that remained in a state of fluidity would have retained a "water" property which would then logically be why we call it "water". Think of this in terms of when we call something "watery".

Keep in mind here that the term "water" itself is not critical, because in Latin it is "aquas" and in German is it "wasser". "waters" is most logically referring to a property that we would think of as fluid or fluidity. As a side note, in Hebrew the letter pronounced as "mem" which in ancient Hebrew is scribed similar to our English letter "M" means water or liquid or fluid. It is the ultimate root of *mammary*, or *mam*, or *mum*, or *mom*. This is only a consideration, but since the term "Maria" is used in Latin for "seas" and contains a Hebrew *mem* and the Hebrew mem basically means fluidity, it seems logical that the term "mem" was the likely root-word for "waters" in the first sentence of Genesis.

Once the H^2O water was gathered, only then could the plants be brought forth. But how did those plants come to be? Did they evolve or were they each specifically created?

What Does the Bible Actually Say?

The question is not whether the plants evolved or were created, but rather the question more specifically is, did it occur as the result of a Creator, or was it the result of random evolution that happened through unlikely circumstances?

If you are in, oh say, your fourth or fifth decade of life or later, you should have regularly been witness to the ability that plants have to live. Plants need little care from us to thrive. We can certainly affect them and guide them, but vegetation in general does not need us humans. The various vegetation types want to live and they do so very quickly. We can reasonably safely assume that this was true during the creation account of Genesis One on day three. However, the problem we are stuck with is the chicken-versus-egg question. What came first, the seed or the tree? Are we to imagine that the Creator fashioned seeds and then planted them?

We have a big problem with science in our modern age where it is believed that somehow the Bible can in no way agree with science, but this simply is not true. What does the Bible actually say about creation? I am specifically referring to what the Bible says rather than **what we think** it says, because the two are typically very different. What we think often differs a great deal from what is written. If it was not, then there would be only one "religion" and one Bible version.

Pay very close attention to the text and what is being indicated regarding the plants. The first key point to consider here is that water, as we know it today, was likely what was called "seas" and it is generally a prerequisite for vegetation to begin.

Here is the pure text of day three alone:

"god also said let the waters that are under the heaven be gathered together into one place and let the dry land appear and it was so done and god called the dry land earth and the gathering together of the waters he called seas and god saw that it was good and he said let the earth bring forth the green herb and such as may seed and the fruit tree yielding fruit after its kind which may have seed in itself upon the earth and it was so done and the earth brought forth the green herb and such as yieldeth seed according to its kind and the tree that beareth fruit having seed each one according to its kind and god saw that it was good and the evening and the morning were the third day"

Read it over a few times very slowly with an open mind and an open heart—some truths will begin to emerge.

Chapter 4

Defining the Argument

Arguments often arise between people of "science" and people of religion regarding the general topic of creation, but also regarding the finer points such as the advent of plant life on Earth. Science, with all of its razor-sharp-bio-speak is a real challenge to the often glossed over POOF!-God-did-it-all religious view. But in truth these two must fully and completely agree or one or both are wrong.

The ultimate *argument* has never been over "How did it occur?" The *argument* has always been, "Did a Creator do it all?" But the *quest* is to understand *how* it occurred. How it occurred is a real sticking point for science because the initialization of plant life is a bit of a stretch from a scientific standpoint, so those who refuse to acknowledge a Creator are then forced to invent a theory for a viable starting point for plant life. This is the part where science tends to gloss over things, much the way science glosses over anomalies in big bang theology by doing things such as nullifying the laws of physics or ignoring the fact that *nothing* was there. Momentarily nullifying the laws of physics allows for the bang to occur, that is to say, physics that are *required* for the pre-bang non-existent energy state that is claimed was there when nothing was there.

On the other side of the argument we have the typical creationists who gloss over reality and assume that God just made all of the plants suddenly appear as we see them today. But is either argument realistic? Is it realistic to assume that the plants developed without any guidance? Or is it realistic to assume that God just said "let there be trees" and then POOF! there were trees everywhere?

In reality, neither approach is honest or scientific. But even from a provable scientific point of view, we are truly only left with speculation based upon currently observable evidence through current plants, the fossil record, human logic, and the Bible's creation account.

The arguments in the public forum are typically spearheaded by the talking heads of pop-science and religion, and truth be told these pop-science and pop-religion icons are not necessarily the foremost authorities on these subjects. Many of them are very knowledgeable, but often refuse to acknowledge *all* of the data found within even their own realm. Objectivity is typically fleeting in such debates, and their perspectives are typically based upon *what they want to believe* rather than on true logic. In religion, logic is often blatantly cast aside and replaced with blind faith. But in science, logic is ignored and *opinion* is then inserted and is renamed as "logic". Neither is useful to Truth. We need to get beyond ourselves and our hardened opinions in religion and in science because the us-versus-them approach typically doesn't advance our true understanding.

Let us define the argument not as an argument, but rather as a quest to understand the origins of all things. If you are reading this book *The Science Of God Volume 2 – Day Three - Gravity, Land, Seas, and Evolution of Plants* it is my hope that you have also read *The Science Of God Volume I – The First Four Days* as there are many points made within it that aid in understanding how the initial creation processes likely began. If we are going to believe that some Creator did it all, then we must consistently follow the text that clued us into a Creator to begin with.

Since, from a Biblical standpoint, we are assuming that a Creator exists, so the quest in this volume is to understand *how* the plants came to be. And since we are basing this on the Genesis One text, we must assume that a Creator was active in the process to a substantial extent. But was that in the form of random evolutionary processes from nothing, or was it an interactive creative process?

Capillary Activity

As the text reads, the Creator did not make the plants directly. The terminology reads that the Creator said "Let the Earth bring forth." This really is a very important statement, and the distinction between the Creator specifically creating the plants and the Earth "bringing forth" is critically important in the analysis of this subject. Creationists typically gloss over this and say that God created the plants, and that's generally the end of the story in their description of things. But it was the "Earth" that "brought forth". Not that the "Earth" itself created, rather this is another one of those organizational turning points. The plants were brought forth, each after their own kind. Here is the pure text version of day three text again with a key part underlined:

"god also said let the waters that are under the heaven be gathered together into one place and let the dry land appear and it was so done and god called the dry land earth and the gathering together of the waters he called seas and god saw that it was good and <u>he said let the earth bring forth the green herb and such as may seed and the fruit tree yielding fruit after its kind which may have seed in itself upon the earth</u> and it was so done and the earth brought forth the green herb and such as yieldeth seed according to its kind and the tree that beareth fruit having seed each one according to its kind and god saw that it was good and the evening and the morning were the third day"

There is a lot more in this brief underlined statement than meets the eye. As we dive into it, we will try to keep it simple and somewhat superficial because many aspects of this entire topic are each lifelong study areas for people, such as with microbiology. Microbiological activity acts as an uncountable amount of micro machines that are created to do *very specific* tasks.

In the highly unlikely event that the Creator just magically made everything appear with a sudden POOF! type creation, there is no sense in even discussing any of this because it simply cannot be figured out in that case. However, since we have been given the gift of God-like logic, and since we have been asked to seek out God, we have to assume that the Creator *wants* us to explore and understand creation.

Once we dispense with the hocus-pocus creation myths and step into a scientific perspective (we are referring to *real* science here, not pop-science), then we can begin to understand the text of Genesis One verse eleven and verse twelve. What actually occurred in verse eleven? Did God make the plants?

No, the Creator gave instruction that the earth should bring forth. "he said let the earth bring forth the green herb and such as may seed and the fruit tree yielding fruit after its kind". There is in no way any sort of time frame stated here, nor is there any reference to fully grown plants or any reference to God creating them directly. However, there is a very broad instruction that the earth should bring forth at least two types of plants, the green herb and the fruit tree, both of which produce seed.

So here we are left with a massive problem with our real-life experience compared with the account of the plants that were brought forth. We know from experience that most plants don't do well without the Sun. And since it is likely that the Sun and moon and stars did not cast any light on the plants until day four according to the Genesis One text, we are left with plants that really could not progress much past the rooting and sprout stages of growth.

Here again we must pause for a moment and rid ourselves of the idea that these were twenty-four-hour days, because "hours" and "days" as we know them today could not have occurred until after the fourth day had been completed with its "for signs and for seasons and for days and years".

This allows for the items that were brought forth by the earth to have had an unlimited duration to germinate and develop. The question is, are they required to have been fully grown into trees or green herbs in the text, or can the rooting and sprouts keep recurring? This needs to be questioned because there is no sunlight at this point.

When we begin to split hairs on this subject it can get very complex, so we will try to keep it simple. Plants use wetting or capillary action to draw nutrient-laden liquid through their cellular structure. Now from an evolutionary perspective, this wetting or capillary action is quite convenient. However, evolutionary processes alone are left wanting because they assume no outside guidance occurred and only a survival of the fittest guidance system is allowed.

As the cells of the plants grouped, they were wetted, as cells are, and when wet items come in contact they sort of pop together, much the way two drops of water will when you take a pin to influence them to join, suddenly they pop into one larger drop. So, from an evolutionary perspective this property of wetting works nicely; the cells can join together forming ever larger structures, eventually allowing the capillary action of the newly formed structure to port water to its various parts. But there is a problem with this, and it is that while cells to cling together, they also divide. How does this cellular division occur from natural processes alone?

Proving Our Hypothesis

When we hypothesize evolution based solely upon uninfluenced activity, meaning that no intelligent force affected it in any way, we must then prove our hypothesis. But can we do so rationally? Creator-less evolution glosses over many problems with its view of unguided evolution. Ardent evolutionists typically say that evolution is guided by the environment, which is true, but that still does not prove that it is done without any creative guidance.

Just because we can hypothesize something and can "prove it" through predictive evidence, it does not mean that our hypothesis is correct. For instance, what happens if we believe that Santa Clause comes just because there are presents under the tree? Does that mean that our hypothesis is correct? Is there really a fat man in a red suit from the North Pole that delivers presents to us? Science is no different. Our explanation does not constitute reality, but rather, reality constitutes our hypothesis. Our problem is our perception of that reality and our willingness to deceive ourselves to serve our own desire in order for our own thoughts to appear to us to be true.

Using Bad Examples

We can go on for days about the inaccuracies of creator-less evolution, but does this give a pass to Christians who espouse Creator-done creation? There is a seemingly clever analogy made by creationists using a watch, which goes something like this: "If you were walking along and found a watch on the ground, you would know that it was made by man and that it was created through intelligence." Is this a fair analogy? While it is clever and true for the most part, it's not a proper analogy for the creation of plants. Plants are alive and they change and grow and they die—a watch does not. A watch has very clear evidence that it was made using intelligence. The same is not true of plants in that same way because a watch is an item that came after mankind, but mankind has never known a world without plants. We can see the clever way that plants are structured, but they will live or die without us, and a watch could obviously not have been there without us.

Does this make the watch analogy invalid and the evolutionist correct? Not really. While the watch analogy is not a good argument for the debate, it does share something with plants, and that is that they are not specifically random. Stepping aside of the potential nothing-is-random argument pertaining to forces of physics, consider this: When something explodes, for instance an asteroid impacts our Earth, we will see a recognizable depression and a random scattering of debris. The scattering is random, but it is also influenced by what we humans have decided to name "the laws of physics". Yet there is little order to the scattering other than the debris distance versus impact force. However, when we look at any organism, we find an incredibly high state of order, which is only increased by its repeatability; which, again, coincides with Genesis One verse eleven. An assembled watch shares that order, but nothing near the magnitude of order as is observed in any plant or plant cell we see. Plants display an enormous amount of order.

Are there better analogies? Maybe, but not much better. The best analogy we can truly give is when in our labs we try to replicate life. We come close to replicating actions of cells, yet they are not true cells, but are impressive. However, so far, we have failed to create anything near a real living and growing life using our human intelligence without using or copying any part of what is already there.

Finding Our Way

When we use floundering analogies, we damage our position regardless of which side we are on in the discussion. As we drill down to the core of this topic to find our way to the truth of the events detailed in Genesis One, we have to get more and more scientifically detailed.

When considering the lab example given in the last section discussing attempts at recreating life in the laboratory, if we can eventually cause a cell to be fully functional through natural processes, then we have proven that evolution could have occurred free of any intelligible influence, right? Not so fast! While impressive as it would be for us to create a single living cell from scratch, we have to make that cell actually do something, such as replicate and form an entity or structure of its own. It still needs instructions, which as far as we can determine comes from DNA.

But, even if we were able to do this from scratch using nothing but natural processes, there is still one looming problem with unguided evolution, and it is that every step of the way in our attempt to create life in a lab from scratch through natural processes not only is a cheap attempt at copying what already exists, but it also <u>uses our intelligence</u> *every step of the way*. There is no getting around this point and no way of ignoring it. You cannot interact with nature as a human being to try to achieve a desired outcome and then tell the world that it occurred on its own, naturally, without human intelligence—To do so is disingenuous and an outright lie.

Chapter 5

A Need for Oxygen

The topic of how plants came to be is truly a complex topic to grapple with because plants require a lot in order for them to thrive. For instance, they generally need oxygen. But since plants create oxygen it gets a bit confusing. Plants take CO^2 and H^2O and refine them in the photosynthesis process producing sugars and oxygen. Each plant is a mini chemical refinery unto itself doing things that we humans would love to be able to do with such efficiency. The question is, when getting into the molecular level, can a plant produce any oxygen without having any oxygen? If a plant could produce a single oxygen molecule and then use that newly assembled oxygen molecule for its aerobic respiration, it could then multiply that process and increase oxygen levels. This sort of activity could have been occurring everywhere on every earth-like planet.

The Burden of Proof

I once watched a debate where one of the debaters stated that the burden of proof is on the person who is making a claim that is not a "natural" claim; meaning that if the claim cannot be proven, or is not evident and easily seen through natural means, then the person making the claims must provide proof. The attitude is that you might as well save your breath and not even bother proposing a theory or making a non-natural claim if you don't have proof.

This is a trap for anyone who buys into this thinking. While there is a point behind this, it defies sound logic. This common view is mostly held by those who disagree with or refuse to accept any thought of Biblical creation. Their point is not to be overlooked though, because it was formed largely due to the ignorance of some past Christians.

When Bible believers make claims that they base solely upon their own interpretation of the Bible and insist that everyone should believe what they say because "the Bible told me so", then we have a real problem in regard to being scientific about it. This is not to say that the Bible is not true or accurate, but rather that the rationale is baseless when people living by the the-bible-told-me-so principle can offer no plausible scientific explanation. So again, there is good reason for non-believers to demand evidence from those believers whenever the-bible-told-me-so principle is invoked.

The attitude where the burden of proof is on the person making "unnatural" claims is actually a good principle that we should all completely agree with. However, the "trap" that I mentioned is not the concept of the burden-of-proof in itself, but rather it is the *application* of the burden-of-proof that becomes the trap. Understandably, those who most frequently invoke this concept are atheists and agnostics. But sadly, too many people who believe the Bible is true and accurate are utterly lost in this regard because they believe their claims without any natural evidence. However, their lack of ability to find and claim the evidence has no bearing on the Bible's actual account of creation; rather, their lack of evidence speaks to their lack of devotion to the subject at hand. If we have rational evidence, then we would surely bring it forth. A *true believer* who truly wants to discuss this topic is not going to overlook the scientific or natural aspects of it.

From a perspective of debate, this idea, which is "the burden of proof lies with the person making claims that are unnatural" should also be used by those who support Biblical creation regarding those who believe in godless-evolution. The proponents of evolution and the big bang are primarily the people using this tactic, but what too many creationists fail to notice is that the big bang and evolution supporters fail to use this argument on their own thinking between their own peers. Much of evolution and big bang theology glosses over that which it cannot explain and for which it is unable to offer true and rational evidentiary proof. Instead you will get "sophisticated scientific" details of specific unproven theories and told things like "scientific consensus is such and such".

The trap that I speak of is not the concept itself, but rather the use of the concept. People who want to believe Biblical creation are often either too lazy, too afraid, or simply do not care enough to do the research or put forth the effort of thought that it takes to properly debate people who believe in godless-evolution and the big bang. "The burden of proof lies with the person making claims that are unnatural" concept applies to **both** sides of the debate. People, who believe that the Bible is true get themselves backed into the corner as they are being bludgeoned with the "the burden of proof" stunt because they fail to return the sentiment. When it comes to creation there is not only evolution versus creation, there are also various views of evolution and various views of creation, most of which have gaps in their explanations even though they claim that there are no gaps. Let us make no mistake about the fact that pop-science's version of "evolution" has more flaws in it than evolutionists pretend that it does not have.

Realize that if this so-called "God" exists in reality, then creationists are ultimately debating the winning argument. Yet, this is not how most promoters of Biblical creation conduct their debates. When you misinterpret the text of Genesis One, you are likely to appear foolish and you are also likely to make God appear foolish. Biblical creation proponents typically take one of two paths in discourse about the Bible. The primary path is that of "I am right and you are wrong because the Bible says so" and the other is to withdraw in shame because they cannot explain why they believe what they believe as they are bludgeoned with the perspective of "the burden of proof lies with the person making claims that are unnatural". Push back the same burden of proof sentiment when debating so as to remove the double standard of having to alone bear proof. What we believe to be scientific proof is not always the proof that we believe it to be.

Insufficient Evidence

The "burden of proof" requires evidence, but what amount of evidence is sufficient?

Since atheists believe that there is no Creator, "creation" is no longer allowed in their minds as any sort of evidence of an intelligent creation. Thus, if they can convince you that it all happened from "natural events", then this is evidence that there is no Creator—in their eyes. But I contend that once each Creation process began when the "spirit of God moved over the waters", then most activity was so-called "natural" from those points on, and thus will be explainable through natural causes other than each specific listed command being issued thus invoking each new action.

We rationalize that we can explain away a Creator because we are able to explain the sequence of events that occurred to produce certain vegetation because it is all "natural". The reason that this philosophy has been able to be perpetuated is because Christians doing Bible-thumping for God want to believe that the Creator practices Santa-Claus-like magic. But the Creator has nothing to hide and therefore all things of the Creator can be known, as I have mentioned in other books. But, just as it is unlikely that anyone can know another person completely because they have not lived that other person's life, so too is it that we cannot know the Creator completely, though we can likely know any one thing about the Creator or the Creator's actions.

Utilizing the conventional pop-science evolutionary approach typically used by the evolutionist side of the topic, the evolution evidence that I have witnessed has thus far always been insufficient as a form of "proof". "Proof" is often given through predictions, but is this predictive "proof" absolute?

The Evidence is Waiting to Be Found

Predictions fulfilled are not always evidence of a specific act. While such predictions can be such evidence, they can also be fabrication that could potentially result in the same outcome that another set of circumstances would produce. In our Santa Claus example, we would first look to see if the prediction is repeatable, and it likely would be especially at the home where they happened to catch a glimpse of this red-fur-clad gift giver. If they were to predict that they might see this character again around the time of December 24th or 25th near or at that home next year, is that then evidence of Santa Claus? Or is it more likely possible that it is evidence of old uncle Bill doing Mom and Dad a favor again by playing a Santa Claus character in effort to fool their now five-year-old?

Is this Santa Claus analogy a fair one? Some may claim that it is not, but if creator**less** all-natural evolution doesn't really exist, then the analogy is a perfect fit! There could be many theories as to how the gifts appear under the Christmas tree, and many of those theories could likely be "proven" with predictive evidence. Yet, the various theories produced might have zero compatibility beyond the connection of the gifts appearing under the tree.

For instance, we could make the prediction that gifts magically appear in the middle of the night. And if we had a time-lapse camera set up to snap a photo every five minutes, we would possibly see gifts magically appearing with no visible cause, thus "proving" that the gifts just magically appear. But if we were to theorize that the parents had a hand in this and we were to follow their every move preceding Christmas Eve, then we would likely see firsthand that they went shopping, bought gifts, brought them home, wrapped them, hid them from the children, and then on Christmas Eve after their children were asleep they would likely be found placing the newly purchased and wrapped gifts under the tree for their children.

Now, we could argue that this is no evidence because it is only an evidentiary prediction, as are the other two predictions. However, this particular prediction and the verification thereof witnesses the actual act of placing the gifts beneath the Christmas tree. Thus, it is the prediction that has the most credibility. However, we still must be open to other possibilities, though once this prediction is known it is not particularly likely that given its accuracy in every aspect that anyone would choose to challenge it. But, if we saw old uncle Bill out and about buying the very same presents as will appear under the tree and we saw old uncle Bill bringing the presents into the house while wearing a red fur coat, then we can safely assume that he is actually Santa Claus by the same standard that pop-science assumes godless-evolution. Yet, old Uncle Bill is still only doing Mom and Dad a favor by picking up the gifts, and he happens to own a red-fur coat. He also only placed them on the kitchen table, so any thought of him placing them under the tree would then be false. Thus, just because connections exist does not mean that our interpretations of those connections are accurate.

If an all-powerful Creator exists and desires for us to discover and proclaim the Creator's existence, then it stands to reason that the Creator would *need* to allow people who are partially wrong and partially right to be able to have their information advanced, regardless of accuracy. It does the Creator no good to not allow anything to be said until we suddenly get it all right all at one time, because then nothing could ever be learned or advanced. Understanding is a progressive activity where we must learn or realize things in our mind first, before we are able to recognize them in the natural world. But we must also realize that sometimes theories need to be changed to adjust to recent additions to information or revelation of errors in our theories.

When looking to the Biblical creation account we can make predictions based upon the text and we should also be able to make predictions of what the Bible would say next based upon nature. For instance, we can predict the order of the Biblical events by considering nature and we can predict nature by considering the Biblical events such as: Oxygen is somewhat of a prerequisite for life, but not necessarily at all stages of life. Some seeds can germinate without oxygen, but will need some amount of moisture and some amount of heat that is substantially above absolute zero (absolute zero is roughly negative 460 degrees below our normal zero degrees Fahrenheit). But in this case, we are discussing seeds, and Genesis One verse eleven says nothing about seeds bringing forth the first generation of plants, but it does indicate that those plants would produce seeds "each yielding seed after its kind".

Based upon our observation of nature we can predict that the Bible's basic creation account should indicate that plants would be "each yielding seed after its kind". And inversely, we can also predict from the Bible that we would find in nature that plants would produce seed each after their own kind. These are predictions that are undeniably evident. Though, this will be scoffed at because someone could look at anything after the fact and write it down and call it a "prediction". But that is not what the Bible is doing. The Bible is not predicting, it is telling what occurred after the fact, but we can make predictions of what we will find based upon the Genesis creation account.

Proof versus Predictions

"Proof" is not necessarily evidence, and evidence is not necessarily proof. I hear and see very little prediction on the part of creationists, but I see a great deal of prediction on the part of science. Often these scientific predictions are touted as "undeniable proof" of their theories and proposals. But using our Santa-Claus example shows that this sort of predicting is nothing more than blind faith in the proposal. If you are told of a jolly old man dressed in red who gives out presents on Christmas Eve, and you then propose a theory that this can be made evident if we find presents under the Christmas Tree, then that would be a prediction which in many homes will in fact be the case–Thus, proving the existence of Santa Claus... to a four-year-old.

But here we have to stop and re-examine the prediction, the fulfilment of that prediction, we have to re-examine the found evidence, and more importantly we also have to consider and examine other possible explanations. This is where science typically fails. Science, and I use the term broadly and recklessly in this situation, will often fail to seek alternate explanation of the anticipated discovery. Let's say that someone proposes that the mother and father in the household have placed the gifts under the Christmas tree and receipts are produced to show that the same items that lay under the tree have recently been purchased by the parents as demonstrated by the record on the receipt.

At this point, science would further seek to make predictions that there is proof of a Santa Claus delivering those gifts and placing them there. So, a group of scientists set up cameras around the city in hopes of catching a glimpse of the jolly old character. After having their cameras recording for several years they finally get their evidence of a fat man dressed in a red-fur coat with white trim who is carrying a bag of toys into a home. This "proves" that Santa Claus exists by typical scientific standards, isn't that correct?

Chapter 6

Does Earth Plus Sun Equal Plants?

As touched on in the last chapter, plants do have a few prerequisites in order to grow and thrive. Some need light, water, and heat, where others can begin without light or much water or heat. In our current experience with plants, they have certain prerequisites. Based on our current experience with growing plants, the seeds are required. Yet, the seed could not have existed until after the plants adequately matured, according to the Genesis One text: "let the earth bring forth the green herb and such as may seed and the fruit tree yielding fruit after its kind which may have seed in itself upon the earth" There is no inference that seed came before plants, and to write any into the text is a perversion of the text. At this point we need to revert to a non-English version of the Genesis one text to seek out clearer detail, especially when referencing "and such".

It is important to understand that translations of any sort cannot escape the fact that the person doing the translation does so with their own biases in their understanding of the text and the specifics of each word being translated. When trying to break down such information into a scientific approach, we have to make sure to not build a house of cards on single words or thoughts in one language that potentially did not specifically exist in the origin language(s). One point to consider between the English and the Latin is that the Latin does not translate to "the green herb and such" as shown in the Douay translation, but rather mentions "vegetation and herb yielding seed" depending upon how it is translated. The King James Version uses the term "grass" rather than "vegetation". So, depending upon how you choose to translate, or which translation book or service you choose to use when translating, the meanings of the words change slightly as the translation source attempts to put each translation request into context.

The more words you feed modern translator utilities the more context they have, but that does not make the translations perfect or accurate. The Latin term "virentem" alone in two different online translators produces "green" in one translator, and "vegetation" in the other. To add to this complexity, we have to consider that the text was originally written in ancient Hebrew. Read the text for yourself in English and in Latin, and if you have the resources to do so then test the words in a translation program with a single word versus translating it in the context of a sentence:

English Douay Genesis 1:11 "and he said let the earth bring forth the green herb <u>and such</u> as may seed and the fruit tree yielding fruit after its kind which may have seed in itself upon the earth and it was so done"

Latin Douay Genesis 1:11 "Et ait : Germinet terra herbam <u>virentem</u>, et facientem semen, et lignum pomiferum faciens fructum juxta genus suum, cujus semen in semetipso sit super terram. Et factum est ita."

English Douay Genesis 1:12 "and the earth brought forth the green herb and such as yieldeth seed according to its kind and the tree that beareth fruit having seed each one according to its kind and god saw that it was good"

Latin Douay Genesis 1:12 "Et protulit terra herbam virentem, et facientem semen juxta genus suum, lignumque faciens fructum, et habens unumquodque sementem secundum speciem suam. Et vidit Deus quod esset bonum"

Many people researching often want use the Masoretic texts as the authoritative source material, but those only date back to around 900 AD, but the results are similar. However, the Vulgate predates the Masoretic texts by about 600 years.

As you can see, using terms like "and such" or adding "grass" or "green" greatly alters the scope of potential description in the verse. Inserting "green" or "grass" is very narrow and disallows all other colors and/or non-grass items. Using the term "and such" is incredibly broad and potentially allows too broad of a definition because "and such" could potentially include rocks. Were "vegetation" covers a wide variety, but only of plants. And "vegetation" cannot legitimately step outside of plant life. In fact it covers all plants, or at least those that would fall under vegetable status. Thus, we have vegetation, herb, and fruit all mentioned in verse eleven which basically covers **all** organic plant life.

Using basic logic, we can deduce that seeds did not come first according to the text in the manner and order in which is it stated. Yet we have to dig a bit deeper to understand the role that sunlight and the earth may have played in the bringing forth of the plants' non-seed beginning.

The Order of Creation

The *order of events* in the account of creation in the Bible's Genesis One text is perhaps the most important aspect to follow if you are going to try to scientifically present the Genesis One text in any discussion.

Looking at the entry of plants, first, of course, it says "let the earth bring forth". So at this point, we could potentially invoke aimless evolution, but since the next statement is a bit more specific, "aimless" is no longer an option. The next statement, "the green herb and such as may seed and the fruit tree yielding fruit after its kind", is another act of separation and organization, similar to the division of the waters where the commands given were "let it divide the waters from the waters" and "let the waters that are under the heaven be gathered together into one place and let the dry land appear".

In some manner, "kinds" where developed in verse eleven. But we must be very careful here in determining the "kinds". And this is where we run into trouble both scientifically and Biblically in understanding the progression of plants. What are the parameters of "kind" stated in the text? And, is a "kind" a type of each "fruit tree" and type of each "green herb and such"? Or does it stop at the division between "herb and such" versus "fruit tree"? Who gets to define "kind" here?

At this point we have to look to our current experience in life and how plants replicate. We know from centuries of writings and from our own experience that plants seed each after their own kind. In other words, an apple does not fall to the ground and then grow into some other random fruit tree, such as an orange tree. And we see a distinct difference between "herb" and "fruit" in our current day understanding.

But we also have to question plants like raspberries and carrots. Do they fit into the green herbs or fruit tree categories? Raspberries and many other plants seem to be overlooked in the text as they are neither green nor trees and are certainly not grass, however they do fit into "and such" or into Latin's "virentem" or vegetation. Is this picking and choosing words from various Bibles to make things fit? Potentially yes, but I am not selecting whichever version suits the topic best, but rather we are going back to some of the oldest available versions to try to understand some of the key points that were intended to be conveyed in the text before the perversion of that text, which took place in the more recent centuries.

Even if we had the absolute original text, we would still likely have to translate it into our modern tongue for us to be able to understand it, and in that alone we could potentially translate the word(s) incorrectly that led us to using terms like "and such", "green", and "grass". If we were so blessed so as to be able to translate from the absolute original text, it still places us in the position of having to attempt to deduce what that original intent might have been. Using the collective of all of the key versions of antiquity of the Bible, we face the same issue. But here we have the collective view of each translator to deduce what was likely intended.

Understanding that the translation issue is and always will be left wanting, we further have to understand "fruit tree". What is "fruit" and what is "tree"? Is a maple tree to be considered a "fruit tree" in the Genesis One account? Or does a maple tree fall into the "green herb" category? And did the earliest texts actually say "tree"?

Look at the English to Latin comparison again:

English Douay Genesis 1:11 "and he said let the earth bring forth the green herb and such as may seed and the fruit tree yielding fruit after its kind which may have seed in itself upon the earth and it was so done"

Latin Douay Genesis 1:11 "Et ait : Germinet terra herbam virentem, et facientem semen, et lignum pomiferum faciens fructum juxta genus suum, cujus semen in semetipso sit super terram. Et factum est ita."

Latin's "lignum" can translate to wood in English and by association wood is from trees, thus we have the translation of "lignum" to "tree". But is this correct? Maybe, maybe not. But it is limiting to a point that the original text may have not intended.

Let's consider the raspberry plant. Raspberries have a stick-like stem structure that is much like wood and it bears fruit. So, by broad definition it is a tree as much as an apple tree is. But there is something more important in the text that we cannot ignore, and that is that the "trees" are not any trees, they are specifically "fruit trees". So here we must further analyze the term "fruit". When we think of fruit, the typical picture in our mind is of apples, oranges, banana, etc, but when we look at the Latin we see the term "fructum" which is similar to our term fructose, which ultimately means fruit sugar.

At this point we begin to get into definitions issues, and since a "Sugar Maple" produces sap from which we can make maple syrup, does that qualify the maple tree as a "fruit tree"? Slicing this fruit a bit thinner "and the fruit tree yielding fruit after its kind which may have seed in itself upon the earth". One computer translation of Latin's "lignum pomiferum faciens fructum juxta genus suum" translates to "and wood pomiferum fruit according to its kind" in English. Notice the term "wood" being used in one computer translation.

Translating is a tough subject because every language has its nuances, and in those nuances words take on slightly different meanings. This makes computers not a particularly good way to translate because it is nearly impossible to account for every language nuance with the translating programs. But one thing that computers typically do not do is use human interpretation influence on single words other than what might have been in the original program. This approach can be used when we must grapple with a word like "tree".

We also have to realize that the culture contemporary to any point of translation might have used a word somewhat differently than we use it today. The Latin "lignum" being translated to "wood" in English is somewhat telling of the omission of specificity that may have been used in the original recording of the Genesis one creation account. There is a possibility that a fibrous stick or stem or vine yielding fruit was being conveyed in the genesis statement of "fruit tree yielding fruit" seen in the Douay version.

Key Bible versions of antiquity agree on the general thought that the seed mentioned in this statement is *in* a part of the fruit itself, which in that case would exclude the sugar maple from the fruit-tree group. Yet, the raspberry then still qualifies as a fruit tree.

There is no indication of size or style, as it only states that these items yield fruit after their kind containing seeds. But that all depends upon what "which may have seed in itself upon the earth" actually means. Does "upon the earth" include all of the plant types mentioned in verse eleven? And does it simply mean that these all occur upon the earth, rather than within the earth? For the time being, we are going to set aside the divisions of the plant types because, from the initiation of plants standpoint, it doesn't matter a great deal.

The Fertile Earth

Most people are familiar with the fact that depleted soil is not very good for plants to grow in, yet some plants will grow in very depleted soil as long as there is some amount of available moisture and a modest amount of heat. We see this occurring after an active volcano flows its lava over large areas. It doesn't take long for plants to begin to spring up on what is essentially porous barren rock. The seeds are likely brought in via bird droppings or by wind currents. If a seed is brought in through bird droppings, the material that the seed is in will have nutrients and moisture enough for the seed to sprout and grow a bit, but other seeds that grow do not have those nutrients available are forced to extract what they can from the lava-rock. Yet plants do grow, and many of those plants are trees whose nutrient needs outstrip those supplied in the bird droppings. Yet a seed can sprout with only a bit of moisture, but it will not get far in growth because the seed has a limited supply of nutrients.

In our life experience, we understand that ground becomes more fertile with the repeated lifecycle of plants rotting and returning to the soil. Eventually the sand or clay becomes darker and rich with bioactivity. Are we then to assume that the earth was created with nutrient rich soil? To an extent, yes.

Depending upon which minerals, natural minerals greatly assist in growth of most vegetation. And since we can witness the bareness of cooled lava-flows and that plants can grow on them, we must then acknowledge that a fresh earth could potentially have many, if not all, of the nutrients needed to start various "kinds" of plants. Not only could plants have had a considerably long period to create their own bed of additional nutrients to grown in, but they might also have been brewing in chemically rich slurry of matter–but for what purpose? Is there a purpose to plants, or are they just something God whipped up for giggles?

Filling a Purpose

While it is likely that it took a great deal of time for the plants to emerge with diversity, we have to wonder if the Creator had a longer-term plan, or if this was all just being thought up on-the-fly as the Creator went along and did creation. If the Creator had no plan, then the plants had no real purpose initially other than to serve themselves by living and dying and fertilizing the ground and making it nutrient rich so that the plants themselves would flourish more with each successive generation, which in itself is a noble plan.

There is one thing that we can all be certain of as witnessed by anyone who has experienced life for more than a couple of decades, and that is that life wants to live voraciously! If we abandon a building and do absolutely no maintenance on it, it does not take long for the building to begin to decay. Once the building's shell is breached, plants will begin to grown on, and in, the building and will eventually cover the entire structure, while forcing walls to break as a tree grows through a crack or a door or a window. Moisture will enter the building making it a literal green-house if it doesn't collapse from decay first. Plants will overcome the building, usually within a few decades, and after a century or two, someone might come across its buried and overcome ruins while clearing trees for excavation as occurs in our modern world from time to time.

Plants fulfill a purpose of enriching themselves and their offspring through their cycle of life. But since on day three the sun could not yet have shone, the plants could not get much past a rooting state. If the plants pierced the surface they would have had no light to begin the photosynthesis process. Therefore, they would likely not have been very green if green at all. All plants could have developed in this state for a very long time.

This plays into the evolutionary perspective nicely, but it still could not have allowed much to evolve because the plants could not do much more than sprout. This is where the instructions would have been developed or given. *Developed* or *given*—one denotes evolution and the other design. Is it one or the other? Or is it maybe a combination of both? Since the account listed in verses eleven and twelve are vague at best, we truly are forced to assume that "vague" is what was intended, and likely so because *vague* is how they were likely created. Since the earth would have been a fresh accumulation of newly generated elements and compounds in random concentrations, likely with differing temperatures, those conditions would have varying impact on the plants that were "brought forth". Each local environment on earth would produce variations of the vegetation, herbs, and fruit plants each producing seed after their own kind.

The term "evolution" has become a divisive term that causes people to take sides because it was hijacked by pop-science for centuries to promote godless origins of all things. However, to "evolve" simply means to change over time with each subsequent generation. So, Creator-based creation had to have included some level of evolution, but it would be limited to only variations of three major "kinds" at minimum. This is something that we have witnessed not only in the geological record, but also with our very own eyes.

Working with what Plants Have

As any botanist knows, you can cross-pollinate many plants of similar type and arrive at a variation of the two plants in the experiment. Many variations have been made by us humans when we cross-pollinate plants, which is something that has been going on for thousands of years.

In science we have another problem similar to the referential reasoning used in the geological column. Definition of species is a topic in itself, but according to Darwin's "species" specifications, which he himself struggled with, he theorized small variances in certain creatures render it a different "species". By the same criteria, Asians, Africans, and Europeans are all different "species". And even though pop-science often claims that there are hard parameters regarding "species", there are no hard parameters on what determines a "species", and the allowed variance is inconsistent from discipline to discipline and from person to person.

In our human understanding, we try to group things into categories, which is a logical approach. But the specific category quantity will depend upon the resolution of our analysis. If we group only by plant type, such as *flower* and *tree*, we have two divisions, but we can subdivide flowers, such as a tulip, into *wild* and *manmade*. And each of those can be divided to somewhere around one-hundred-fifty "species" which can then be further divided into several thousand varieties. There is nothing wrong with our current human definitions of genus, species, and variations etc., but those modern definitions cannot possibly apply to verses eleven and twelve of Genesis one because those are manmade and are relatively modern terms.

What we do know from our scientific experimentation is that some things will successfully crossbreed and some things will not. The closer they are in their structure then the more likely it is that they are compatible. We also know that plants that produce seed will render plants like themselves. We see this proven trillions of times every year.

Evolution is extremely fast as is witnessed when crossbreeding two compatible plants and making a variation of the two; yet it is unlike the two in many ways. We can do this in a matter of a few weeks with fast growing plants; yet, our grass is always our grass, and the trees that spring up from the maple tree in the yard are always maple trees.

In using our current experience coupled with the information in Genesis One chapters eleven and twelve, we can confirm the production of subsequent plants "according to its kind". And through the replication of plants and environmental changes, we can also witness variations within kinds. This can be witnessed in vineyards by grapes producing varying flavor based upon what chemistry is obtained from the soil, that is to say nutrients, and the type of soil they grow in will affect their texture and taste.

These changes happen quickly and cause a tremendous amount of variety. What we do *not* see is any deviation from major kinds. And we have no evidence of change of major kinds. In fact, the fossil record shows little or nothing and then suddenly there is an explosion of life of many varieties.

Our real-life experience stands in testament to every word of Genesis one verses eleven and twelve, and we find nothing to scientifically defy this. In fact, it is much the opposite of pop-science's view because the description in Genesis One, while brief and to the point, is precisely what we see in nature.

Chapter 7

The Protective Earth

From our human quest to explore space we have discovered, without question, that space is a hostile environment that is not particularly friendly to life. Space is cold and in itself is uninhabitable. There is no oxygen or water or heat, but radiation abounds and it can kill in relatively short order. Yet, here we all are alive, right here on Earth, as it sails silently around the Sun year after year, century after century.

Our atmosphere protects us from the hostile environment of space all while we speed through space at nearly seventy-thousand miles per hour around our star, the Sun. It's a pretty cool spaceship! When you consider how hard we try, but mostly fail, to replicate our self-sufficient environment—even within our environment—it is really quite impressive that our Earth does this without effort, something we have yet to have been able to achieve to any reasonable scale or level in our quest to inhabit space.

Within the Womb of Earth

Our atmosphere is much like the womb of a mother. Yet earth has no soul like us humans. Earth has graciously, patiently, and gently carried us humans safely through space, around the sun at least six thousand times during recorded human history, and maybe more, much more depending upon which scientific discipline you ask. In this six-thousand years' timeframe I only speak of time as recorded by humans in the Bible. This excludes any time that we humans were not on this Earth, such as Genesis' days one through five.

On the third day "god also said let the waters that are under the heaven be gathered together into one place and let the dry land appear". There are many people who want to force this into the constraints of, specifically, the seas and/or oceans being somehow separated from the dirt. I am unsure of what may be being pictured in that perspective, but it is a very local and specific perspective. If you had the opportunity to read *The Science Of God Volume I – The First Four Days* you will recall that unless we take extreme liberties with the text, that the "waters" spoken of in the first couple of sentences of Genesis One cannot be water (actual H^{2}O) as we know it today.

Here is the pure text of the first four days for your review "in the beginning god created heaven and earth and the earth was void and empty and darkness was upon the face of the deep and the spirit of god moved over the waters and god said be light made and light was made and god saw the light that it was good and he divided the light from the darkness and he called the light day and the darkness night and there was evening and morning one day and god said let there be a firmament made amidst the waters and let it divide the waters from the waters and god made a firmament and divided the waters that were under the firmament from those that were above the firmament and it was so and god called the firmament heaven and the evening and morning were the second day god also said let the waters that are under the heaven be gathered together into one place and let the dry land appear and it was so done and god called the dry land earth and the gathering together of the waters he called seas and god saw that it was good and he said let the earth bring forth the green herb and such as may seed and the fruit tree yielding fruit after its kind which may have seed in itself upon the earth and it was so done and the earth brought forth the green herb and such as yieldeth seed according to its kind and the tree that beareth fruit having seed each one according to its kind and god saw that it was good and the evening and the morning were the third day and god said let there be lights made in the firmament of heaven to divide the day and the night and let them be for signs and for seasons and for days and years to shine in the firmament of heaven and to give light upon the earth and it was so done and god made two great lights a greater light to rule the day and a lesser light to rule the night and the stars and he set them in the firmament of heaven to shine upon the earth and to rule the day and the night and to divide the light and the darkness and god saw that it was good and the evening and morning were the fourth day"

You will notice if you follow the instances of "waters" underlined in the text, that in no place is it stated that the "waters" is specifically created. Some Bibles take excessive liberties with the term "firmament" by changing it to "expanse", "canopy", "vault", "sky", or even "horizon" ultimately completely altering the scope of the creation text. We'll get deeper into the "waters" issue in a later chapter. But read the pure text of the first four days just shown and see if you yourself can read it and truthfully rationalize that the first separation of "waters" on day two were clouds as the "waters above" and seas as the "waters below". If you are able to ascertain *clouds* and *seas* from the text then you are writing in changes that simply cannot be and were never written in the oldest texts.

The Genesis One text is really much more broad than our narrow-minded earthy perspective allows. Our little mental boxes attempt to confine the creation text and the Creator to our human mental capacity and to our earthly view. But if you can separate your mind from our typical Earth-centric-view and read the text from an all-encompassing overview that the Creator would have, you will then find the text to be far more accurate in all respects. Most of the creation text has little to do with our specific Earth.

On day three as the waters were gathered, it was likely not specifically speaking of H^2O, at least not until the seas were named. The gathering of the earth matter having a "waters" property (something that can flow or be poured) is far more likely to be accurate on a sub-atomic level thus allowing hard elements to be formed, such as atoms, which would further allow compounds to form. This would allow "dry land appear" and "the gathering together of the waters he called seas". At this point the gathering process would still be at the atomic level and would likely be the process of the compounding of atoms into many compounds such as H^2O thus creating "seas".

The compounding of atomic elements would not be exclusive to H^2O in the oceans, but would include all elements including those in the compounds that comprise "dry land" and those that comprise an atmosphere. A suitable atmosphere was likely largely developed during the day three events. And while this early third day atmosphere would likely have been hostile to us humans and not particularly breathable, many new compounds forming would allow the potential for the generation of heat, not only from movement and atomic activity, but also from chemical interactions and reactions of some of the newly formed compounds.

Such an atmosphere would have initially insulated the celestial bodies from the cold of space and allowed for sufficient heat for the earth to bring forth plants. A womb-like atmosphere would have encompassed earth diverse with compounds that would allow initial aspects of plant growth to easily occur. At this point in the discussion we typically run into the false premise that all of the heavier elements are only formed in stars, but that is only speculation because if a single atom of hydrogen is able to form of its own accord then why not more complex atoms? We cannot rid ourselves of our logical anomalies by time-slicing big bang or by minimizing the periodic table down to hydrogen and helium as the sole starting atoms.

Enclosures of a Seed

Our human experience is really quite narrow, and for the most part, we have no way of avoiding this. We are born and we open our eyes and begin to learn. We learn what we experience and it is only when we are a few years old that we really begin to play using our imaginations. But as we age we become more and more accustomed to the way things are in our environment which causes us to stop imagining what else might be. Instead, as adults, we only think of what is within our own environment and our limited adult imaginations, and it is by those parameters that we think and attempt to reason.

So as adults, we eat an apple or maybe a cherry and we find seeds. We are told that if we put those seeds in the ground that then those seeds will grow into a tree and bear fruit like the fruit that the seed was from. We can then test this ourselves, and if we are patient enough to wait about ten years, we will find this to be true. Plants come from seeds in our experience and we have no need to doubt this.

But what is a seed? A seed generally has a protective shell with material inside of it that is the raw material needed for that particular seed to grow into a plant like the plant that it was from or of its kind. Some seeds are large, like the pit in a peach. And some seeds are very small like a sesame seed. Seeds are much more than the little hard shells that we see them as that magically produce plants. Seeds are complex items with a great deal of chemical compounds that are pre-packaged by the plant that the seed came from. This in itself is very amazing, but we have to realize that seeds were likely not specifically a part of the day three events in that they likely were not planted like we do with seeds today because "let the earth bring forth the green herb and such <u>as may seed</u>".

Abiding In Harmony

While the day three events do mention "seeds", we need to further examine the text. The first point to consider is that day three says "let the earth bring forth the green herb and such as may seed and the fruit tree yielding fruit after its kind which may have seed in itself upon the earth and it was so done and the earth brought forth the green herb and such as yieldeth seed according to its kind and the tree that beareth fruit having seed each one according to its kind". This does not require that seeds as we know them today be fully formed. Take a look at the Latin version:

English Douay Genesis 1:11 "and he said let the earth bring forth the green herb and such as may <u>seed</u> and the fruit tree yielding fruit after its kind which may have seed in itself upon the earth and it was so done"

Latin Douay Genesis 1:11 "Et ait : Germinet terra herbam virentem, et facientem <u>semen</u>, et lignum pomiferum faciens fructum juxta genus suum, cujus semen in semetipso sit super terram. Et factum est ita."

Take note of the word "semen" being used. In our human experience a seed is a complex structure that results from matured plants. But is the "seed" or the "semen" mentioned in the day three events actually a seed as we know it today, or is it something more critical?

The text says "and it was so done and the earth brought forth the green herb and such as yieldeth seed according to its kind and the tree that beareth fruit having seed each one according to its kind". It does **not** say that these plants fully grew at that point, but rather that the plants were brought forth and had seed. Further, we must recognize that the term "green grass" or "and such" used in some translations is more likely intended to be what we understand as general vegetation, rather than green anything and thus the photosynthesis that occurs due to light would not have been required for the day three events to occur.

Day three is still timeless because there could not have been any light shining upon the earth until day four when the lights were put into action. This would allow the potential plants a great deal of time to abide in the earth, rooting and re-rooting time and time again. But how would they re-root without seeds fully forming and re-seeding the ground? The likelihood that the "seed" or "semen" referred to in verses eleven and twelve are not matured seeds is very high. It is almost certain to be partially a reference to what we today call "DNA" or the specific instructions that separated the "kinds" stated in the text.

As the newly "brought forth" plants abode, each in the environment local to them, they would have adapted to varying conditions and would have developed in a manner that would allow them to vary but would be constrained to "according to its kind". This is obvious in our environment today when we take root shoots from plants. It is the DNA, or the instructions, for "each after its kind" that we call DNA. This allows for variances and yet constrains plants to the broader kinds. As the plants abode on earth in their chemically rich environments, they would have been able to develop each their own unique characteristics and would have had ample time to do so since "days" could not be counted because the sun did not yet shine

They Found Their Home

As we can witness today by taking a piece of root from a plant, when done properly, the root will produce plants of like kind, so it's not true that we need fully formed seeds to make more plants when we know how to do it properly. This is done world round and has been done for centuries if not millennia by us humans. When the earth was commanded to "bring forth", it is at this point that the DNA would have begun to form or was instructed. I see no specific explanation in the text how this may have occurred other than, based upon our current contemporary knowledge, DNA is the determining substance that directs the chemical compounds and elements to be guided into the various forms that was most likely being referred to by the use of the term "seed", or "semen".

The plants had found their home in the earth and would have had no time constraints regarding how long they had to abide in the earth before they became similar to the form in which we recognize them today.

Our human minds tend to want to package everything into recognizable little boxes that we are comfortable with. But the scope of the Genesis One creation text is likely far outside of the comfort of our little mental boxes that do not allow for our reading of "seeds" or "semen" to mean DNA. In fact, I would be willing to wager that when the term "semen" was pointed out, the thought in your mind was likely that of humans. Was it not? And further, when the term "seed" is read, you are likely picturing any seed you might be familiar with, such as an apple seed or a pumpkin seed or a watermelon seed. Was it not?

As you can see, it is our limited human perspective that stops us from understanding the creation text with the richness with which it is likely intended. Root systems do not need a sun/star to get started, they just need chemicals, heat, and a bit of guidance to get under way. We can easily, without stretching or forcing any of the text, come to understand that the chemicals and the needed heat would likely have been available on day three. But the DNA instructions, on the other hand, are a bit of a different story.

Did the plants all evolve from nothing, without any guidance? Or were they guided in some way? From an evolutionary point of view, the newly created earth environment was prime for evolution to occur, but to what extent? And the "to what extent" part has always been the dividing factor between godless-evolution and God-based-creation.

To what extent did the plants evolve? From our earthly contemporary experience and the writings of people from past human history, we can be reasonably certain that there are divisions of types of plants that on their own do not converge to create new variations, which supports the "each after their own kind" statements. We, as humans, can force certain mutations between kinds, but those generally do not live or thrive in any meaningful way. Yet, when we merge plants within a kind they often live and sometimes they thrive and are even very beautiful.

The day three text is full of information that our little mental boxes shield us from seeing, and until we smash those little boxes we will remain trapped in them unable to advance our understanding of the text that has so meticulously been passed on to us through thousands of years.

We could think that God should have just said DNA instead of "seed" or "semen" or maybe used some other familiar indications. But then we would have to question what DNA was. And if the text said DeoxyriboNucleic Acid, then what would we have understood that as? The details at that point start an avalanche of information that most of us are not mentally prepared for. Nor do most people want or need that much information. Additionally, the term DeoxyriboNucleic Acid is a man-made term that the Creator likely did not imagine uttering using those sounds or symbols.

In truth, the terms "seed" or "semen" are both sufficient in a broad sense, and it is only our narrow human perception that hinders us from understanding the very wide scope of view that is embodied in the Genesis One creation text. It is very possible that the seed or semen mentioned in Genesis One was meant as DNA, but we see it as the whole seed, shell and all, rather than the DNA contained within its contents. We truly need to reevaluate the day-three text to better understand the entirety of Genesis One.

Chapter 8

A Time before Sun

An area of great debate and confusion is how the plants could be "brought forth" *before* the Sun ever shone upon them and the Earth. This is a sticking point for some creationists and one of the reasons that six-twenty-four-hour-day view is accepted by too many of them. If we imagine that on Tuesday, the third day of the week, that the waters were gathered and the plants were brought forth, then we only have to imagine that on Wednesday, the very next day, or the fourth day of the week, that the Sun was shining in the morning, much like in our current day-to-day experience. If we gloss over the technicalities of this, then this is comfortable and works nicely in our human minds. But it ignores almost every other aspect of the text that we have been evaluating and is utterly unrealistic and very unscientific.

How could the plants grow without the sun, since the sun seems to be a prerequisite for plants to thrive based on our life experience? We discussed this a bit in the last chapter, but let's look at it from another approach to see if the last chapter is grasping in order to force-rationalize it into a scientific perspective, or if it can stand against other events that could have or would have been required to occur in harmony with what was pointed out in the last chapter.

While we discussed DNA in the last chapter we did not speak of the broader part that contains the DNA—the cell.

Dividing Cells

We can talk about plants being in their embryonic state—"Embryonic" has a nice ring to it and we can leave it at that because embryonic sounds comfortable and relatable, but it glosses over a great deal. What is the "embryonic state" of a plant? Is it the DNA? Is it the seeds as we know them today? Here we get into human terms again, much like DNA or **D**eoxyribo-**N**ucleic **A**cid. *Embryo* is another term that requires definition, but since it is not in the Genesis One text, it is mostly irrelevant to this entire discussion, except that its underlying meaning has a great deal of relevance. And it is the underlying meaning that is important to the discussion of the creation of the plants.

The first point to understand here is that, as mentioned in a previous chapter, the plants were not specifically made or created directly by the Creator. This is an incredibly important point to grasp when trying to define the debate between evolution and creation, and also in trying to find the proper approach to the topic. Now and in the past century, the six-twenty-four-hour-days creationists have trapped themselves in the fight against evolution by fighting the *term* "evolution" only because the term was hijacked by godless-evolutionists.

The godless evolutionists pushed the evolution issue to a point beyond its actual scope and they would then, and still do, gloss over the finer points, thus allowing their theories to work in their own eyes; and sadly, in the eyes of some six-twenty-four-hour-day creationists. This causes those who refuse to abandoned incorrect or inaccurate beliefs to set up a fortress of God-did-it-all-because-the-Bible-told-me-so, which is really potentially very accurate, but at the same time very foolish.

When we fight to defend our place, regardless of which side we argue, we had better check our information compared to the information of those we are debating, or we are likely to be made to look foolish. It really doesn't matter which side you take in the debate, because the same is true regardless.

In the case of plants, both evolution and creation are accurate, and it cannot be one or the other. We can clearly see how plants change from generation to generation, so from that we can understand that evolution exists, but to state that "God did it" and leave it at that, causes you to lose credibility in the debate because you offer no rationale for it other than "the Bible told me so". This causes observers of the debate to flee from the Bible because they have nothing solid to base their belief on, other than the slippery slope of the glossed over text.

According to the Bible, it is true that "God created", but what was "created" and what was "brought forth"? This debate cannot be properly framed until we are able to differentiate these seemingly little nuances in the text. And further, since what we read today is a translation, we simply cannot rely only upon a single source Bible. We must include any Bible version of antiquity that we can find, and then use those versions in our analysis of the text. Sometimes when two people attempt to explain something, they each will use different words, but the recipient cannot grasp the meaning without both explanations. Understanding the Bible works the same way because it has been translated and the original copies are no longer available or are hidden from us. This leaves us to forage through multiple versions to ascertain truth of *the original intent*.

The DNA mentioned in the last chapter would likely not have been lying open on the ground newly created. All of our evidence points to the DNA being contained within a cellular structure. We can be reasonably certain that cells had developed after the naming of the seas because H^2O is an important part of cells. And further, cells as far as we can tell, are what all living organisms are constructed of.

People have been doing some very clever experiments in attempt to create cells, but as of the writing of this book they have not successfully done so to any reasonable level. This does not prove anything, but it does demonstrate that the liquids and chemicals that were potentially there when the plants were "brought forth" could have formed rudimentary cellular structures on their own. But those structures would have likely lacked the needed DNA instructions that make them do what cells do. All of this could have easily occurred in an environment that lacked solar energy, provided that the materials were warm enough to flow and move and connect through surface tension and capillary activity.

In our earthly experience, we have taken single cells in the laboratory and manipulated them, and we have been able to grow things from single cells. So it's not a stretch of the mind to assume that the same sort of activity could easily have occurred while the plants were "brought forth". Further, we can contend that much the way two drops of liquid will snap into one larger drop, it is also true that if the surface tension becomes too great, then the liquid can again separate into smaller drops. However, cells do behave differently than the effect that surface tension alone produces.

Multiplying Cells

Every living thing that we see as we walk through our world day by day is made up of cells. But what are the cells themselves made up of? And how do they multiply?

Through surface tension and capillary activity, we can vaguely understand how cells could possibly join, but that could only explain their ability to stick together and only very roughly so. Cells are far more complex than a couple of drops of water. In fact, even a couple of drops of water are far more complex than a couple of drops of water appear to be. Cells, while seemingly liquid or damp in composition, have highly complex activity that causes them to do their function. That function appears to us currently to be dictated by the DNA and the RNA (**R**ibo**N**ucleic **A**cid).

Again, the amount of time that the earth brought forth plants is not specifically stated in verses eleven and twelve of the Genesis One text. And a "day" could not be a *day* as measured by our current method because the Sun could not yet have shone upon earth until day four, thus there is no time limit for the cellular activity to begin and develop. The cause of the separation of cells is not really known. We can sort of explain things in our glossed over way scientifically, but we fail to be able detail why or how this actually occurs. Our best guess is that it is somehow the DNA instruction through RNA that causes this to occur. But this presents a huge problem for godless-evolution.

Us versus Them

There are far too many of us that have trapped ourselves in the us-versus-them mentality, and it does not matter which side of the creation-versus-evolution debate we are on. The contention should never be between "us and them" but rather between truth and not truth.

We often believe that because we discovered DNA that it proves evolution, but this is both true and not true. Evolution certainly exists and is seen regularly in our day-to-day lives as mentioned in an earlier section. However, the critical issue is the *scope* of that evolution, and whether or not it is guided. We know that evolution is guided by environment to some extent, but to what extent is it guided by environment? And is evolution at all guided by intelligent activity in any way?

This is where godless-evolution tends to run off the rails a bit. We humans can define things in whatever framework we choose to define them, but that will not make our definitions correct when they are wrong. Take the idea of DNA for example, what is DNA? DNA is a sequence. It is a program, much like a computer program, but simpler and yet far more complex. As far as we scientifically understand at this point in time, there are four codes used to program a strand of DNA and it is this program in the DNA that instructs the cellular activity. This seems simple enough as if it could occur naturally, until we consider the concept of instruction itself. Instructions are a form of separation or organization much like when "god made a firmament and divided the waters" on day two. Separations are simple actions that cause a great deal of diversity. The earth was instructed to bring forth and it was then that the instruction was likely installed for *kinds* to be developed. We have little more information than that on this subject. Our science tends to agree with this assessment, in that we simply have no real explanation as to how or why cells do what they do other than that the DNA appears to be a semi-programmable instruction set that the cells abide by.

It is not as simple as us against them, because there are as many different viewpoints as there are people on earth, and we are often at odds even with those who "believe as we do". This is true in big bang theology and in creation beliefs, as well as in godless evolution beliefs. Our quest is now and always has been to discover truths. And the discovery of truths cannot occur without *science* or the *creation account* as is made apparent when viewing both with open eyes.

Separate the Issues

This topic gets complicated by our inability to separate issues properly. Most of us don't even know the difference between deist, theist, and Christian. Why would I even bring that up in a book that is mostly about plant development and gravity? I mention it because too often history revisionism occurs in attempt to take the religious devotion and remove it from the scientists of old. Removing their dedication to their God from the picture, repaints them as atheists, which most were certainly **not**. For instance, a person who does not believe that the Christ was the messiah is not necessarily an atheist. Consider the Jewish faith, they do not accept "Jesus" as the promised Savior, thus they are not defined as "Christians". However, they are still theists or deists depending upon each their own personal understanding and belief.

I also mention this because history revisionism tends to cause those who have little information on past scientists to believe that those past scientists did not believe in God or in a deliberate Creator, which is largely not true. In fact, most of the scientists up until the 1900s often had very strong faith and a belief in a Creator. Much of their work was focused on trying to explain creation and how the Creator did things.

Separation of issues is a very important part of any field of study. If you are unable to separate issues or aspects of the topic, then you will tend to blur lines that should not be blurred. It is these blurry gray areas that we typically gloss over. There are few people that will *completely* reject all of the points being made throughout this book series about creation and other prominent Biblical events. And while they might not accept it as fact, they will be forced to consider any part that they have not heard before. Some parts may be known and some parts might be new information to any one person.

This *separating* problem already begins during a glossed over reading of Genesis One. Just the first basic point of the creation of light, versus *when* specifically the sun could have been made causes confusion and glossing for many readers. There are many people who assume that the creation of light is the creation of the Sun, and then by the time we get to the part where the Sun was likely put in place, we have tangled our mind to a point that we have no choice but to gloss over the information that we have convoluted in our mind. This glossing problem then causes us to review the text from a haphazard approach, and thus we say that the ball of dirt we live on that we call "Earth" was created in the very first sentence of Genesis chapter one, which it obviously was not if we are being accurate with interpreting that text and are not rearranging it in our minds.

Broadening our scope of view allows us to see the bigger picture. The Creation account is not about our earth and sun alone. It is about the *entirety* of the "Heavens". And what occurred here on our 'Earth" was most likely occurring on nearly all other planets simultaneously as they orbited their stars. But where conditions did not warrant, those conditions would have inhibited the *bringing forth of*. And the proximity of the planets to their star(s) would also cause extreme environmental variances. Chemistry would be altered by those variables and would cause some of them to have their chemistry be heavily impacted. On some planets, the chemistry might be in a static state unable to react due to extremely low temperatures, where on other planets the environment might be far too hot causing some of the chemistry to burn or boil off, leaving a very hostile environment completely inhospitable to any forms of life that are recognizable to us.

When discussing evolution, there is the *micro* versus *macro* aspects of the topic, which is basically the scope of evolution that was mentioned earlier. But we also have to consider the evolution of microbiology at the cellular level that was just mentioned. If we chase the evolution rabbit down the rabbit hole, we will always end up at the same place, which is, "how did it all get there?" This brings us back to the topic of the big bang which is a better fit for, and is discussed in, *The Science Of God Volume I – The First Four Days*, but the question will always remain, **how did it get there**? How did that first point of bang become, that speck of nothingness that is claimed to have held all of the energy of the universe and to have suddenly and without any intervention banged in order to make all that we see today? **How did it get there**?

This unavoidable point is the single point of evidence that some form of intelligence was first and before all, and this intelligence had duration endless prior to anything in order to accomplish the task of causing nothing to become something. All other explanation will always place the what-came-before carrot-on-the-string, and you will not ever be able to explain it away.

We must separate things like when the "waters" were mentioned, what specifically was really occurring? Were those the same "waters" on day two and on day three? And, was the "waters" on day two the same "waters" that were being "moved over" in the first few statements of Genesis One? Was the "light" made on day one the same thing as the *light* that we see today? Was each "day" in the creation account the same as the 24-hour days that we experience here on Earth today?

In the account of creation in Genesis One, *separation* or *organization* are the single most important action within the entirety of the text. Every statement states some sort of separation throughout the whole of the text. If the reality is that a Creator bestowed this knowledge upon us via the creation account in Genesis One, then we must learn from that account, and learn the importance of that separation, because it is the separation and organization that causes everything to occur. We use this scientific separation every time we use any device with any sort of microchip. Microchips are generally made up of a complex series of on/off switches re-routing electricity to accomplish a planned outcome or state of the switch separators. If we expect to be able to discern the events and the details therein without learning to properly separate the issues, then we are blinded and should expect little or no ability to advance in truly discovering the truth. But what is the truth and who has the authority of truth?

Should Church and State Be Separate

What is truth and who should be the keeper of truth? In the past, the Church, which is to say the followers of God were historically who did what we today refer to as "science". Truth is something that apparently is not understood by far too many of us today. Either that or our desire to not be wrong overshadows *truth* in us, to a point where we completely ignore it so as to rationalize our own erred belief and behavior. Truth is an interesting topic. Even Pontius Pilot asked "What is truth?" when questioning the Christ. "Truth" is a gating system much like the switching in a computer's on/off methods or more to the point *true* or *false*.

Evolutionists typically advocate for separation of *church* and *state* and selectively invoke Thomas Jefferson's personal memos and letters, but they never ask why he wrote or discussed those things. And further, they ignore the other signers of the Constitution, most of whom were quite dedicated in their respective faiths.

Yes, church and state should be separate, but this should not be a discriminatory separation. Yet with each passing year, atheists have cleverly further forced the church out of nearly every aspect of matters that in any way involve the government and science. This is a very peculiar thing since the United States of America was founded on Biblical and Christian principles from top to bottom—You need not look far to find that this is true. The government should never force people to worship in a way that has been defined by the government, because if that government should ever be overtaken by ungodly forces thus placing religion under control of the government, then the people should readily anticipate that their lives are in danger if they do not comply with the government's church. This is why separation of church and state was initially instituted.

Yet, this should never stop the government from promoting things of God on a voluntary basis. Science and godless-evolution receive enormous amounts of money to research those perspectives. Yet, if a Christian organization receives public funds for their research, then it is typically challenged in a court of law as "unconstitutional". This however, is a very hypocritical situation, since godless-evolution is more of a belief system than is Creation.

The biggest problem is that the some post-reformation Bible translations and modernization of Biblical text done in the 1800s has perverted the words of the Genesis One creation account text, which has blinded many readers of those translations, causing a massive rift between science and religion that does not really exist—except for in the hearts of those who foolishly adhere to those perverted versions—which includes atheists. While atheists might not believe the creation account as stated in those particular Bible versions, they do nonetheless believe those Bibles speak for this God in which they proclaim to not believe.

Yes, church and state must remain separate, but state also must *not* utterly exclude religion. Excluding religion entirely is dishonest to the people who pay the taxes, and doing so is insulting to the Framers of the Constitution and to all of the people of the country since that framing.

Chapter 9

Protected Seeds

The "seed" bore by each plant that was brought forth on day three was likely not the same as the seeds we use today. And until the Sun could shine upon the plants on day four the plants would likely not have been able to develop much beyond a small and likely white sprout (etiolate) sticking out of the ground. Based upon our experience with plants, this is generally what we see occur when light is absent.

The "seed" mentioned in verses eleven and twelve were most likely ultimately referring to the plants' DNA and would later eventually develop into the encapsulated protected vessels we refer to as "seeds' in our modern terminology, as the plants produced them.

The Bosom of Earth

We are faced with a monumental challenge today in refuting conventional thinking. Much of that has to do the proliferation of printing over the past several hundred years and has been exacerbated by radio, TV, computers, and now the internet. Conventional modern godless-scientific thinking is that *matter* was somehow gathered into what we call "stars" and through a combination of nuclear fusion and fission, the base element that we know as "hydrogen" was theoretically gathered in such massive quantities that the hydrogen atoms were combined through nuclear fusion forming the more complex atoms we see in our periodic table of elements today.

This is a fair enough assessment of what might have occurred. But is it realistic to imagine that everything was first gathered as stars that were made only of hydrogen and then that hydrogen fused itself together and through that fusion formed the other atomic elements, and then eventually the star exploded and the dust from those explosions gathered and subsequently formed the planets and moons etc? Is this realistic? Possibly.

I do not specifically debate that this over-simplified description is what possibly occurred, but we must ask: How did the first most basic element of the hydrogen atom come to be to begin with? And if the modern contemporary assessment is correct that only a few types of elements initially formed, then why not all of them? How could hydrogen or other low count elements form to begin with? We can use the conventional pop-science explanation and make a compelling argument, but that argument still fails to adequately explain how hydrogen first formed. If hydrogen was able to somehow form of its own accord, then why not also more complex atoms?

Embracing of Earth

One of the problems that science is faced with is the understanding of gravity. We really do not understand gravity the way we need to in order to understand how things came to be. Again, here we are not so much concerned with if it was intelligently created, since we are discussing Genesis One's Creator-based account, but rather we are trying to understand *how* it may have been done. In truth, it is unlikely that we will ever know for sure in our human lifetime on Earth, because all we can do is to offer up possible events that could potentially have caused the elements to assemble. Gravity is the one point that has the most importance in the creation account regarding the assembly of matter. When might gravity have first occurred? It was probably not in the first sentence of Genesis One, but it may have been on day one's introduction of light. Or it could have been when "god said let there be a firmament made amidst the waters and let it divide the waters from the waters" on day two. The "firmament" is a logical place in the text to assume that gravity was initiated because there was a separation of the waters at that point and then the very next activity was "god also said let the waters that are under the heaven be gathered together into one place and let the dry land appear". The first event on day three is definitely indicative of gravity as it introduces the concept of attraction when the waters were "gathered".

Our human understanding of gravity is as follows: *gravitational force, electromagnetic force, weak nuclear force, and strong nuclear force*. And it appears that this attractive nature of matter is key to everything that is seen. If the attractive forces just mentioned were to suddenly cease, then everything would dissolve into nothingness. Gravity, which is to say attractive forces, likely began being formed on day one with the introduction of light and may have been enhanced or added to on days two and three. The day two events are almost certain to be point where a distinction between our tangible universe and that of the Heaven of Angels occurred. The statement, "divided the waters that were under the firmament from those that were above the firmament", is far more insightful than most of us give it credit for.

All too often, I hear people describing this as a separation of clouds and oceans, but that is really not a possibility because the Sun is a big factor in the production of the clouds and in the evaporation of the water. Additionally, the waters could only be first gathered into "seas" on day three. If we are going to take the

Genesis account of creation as a credible source of information then we must respect the categorizing by "day" *segments* or "day" *events*. We must analyze them in the order that they are given without retroactively assigning day three and four events to day one and day two events. Holding firm to the order of events makes it impossible for "divided the waters" mentioned on day two to have anything to do with *clouds* in the *sky* or *water* in the *seas* that we are familiar with today because those were specifically day three events, and they were possibly not really even fully developed at the beginning of day three.

Day three's account of the gathering is the likely point where the formation of the elements took place, and it was also likely a fairly long stretch of time when described using our current Earth years. We must realize that the "waters" as used in the text up to the point where the seas were named is most likely a descriptive property of general *matter*, indicating that the matter was not yet solidified in any tangible way until that event was completed or substantially under way.

On day two we see "god made a firmament and divided the waters that were under the firmament from those that were above" and then the "waters above" are no longer mentioned as "waters above". The next event is on day three when "god also said let the waters that are under the heaven be gathered together into one place and let the dry land appear and it was so done and god called the dry land earth and the gathering together of the waters he called seas and god saw that it was good". This particular event is key in science and lacks a great deal of detail, yet it is the most pivotal point in our scientific understanding of the heavens. It likely includes the assembling of most atomic structures and the subsequent gravity produced by those structures. This would allow for compounds such as minerals and water to become tangible concepts and would then eventually allow "dry land" to appear.

We barely understand light, but we do have a reasonably good grasp on the tangible elements we see in the periodic table, and it is the first part of the third day text's events that allows for those elements to have been assembled. "god also said let the waters that are under the heaven be gathered together into one place". "one place" could be a specific location, but it is more likely a causation of movement and fluidity. Now instead of matter moving without direction, it was gathered or grouped forming compounds or molecules and no longer was able to move randomly apart from each other. It was gathered to the specific location of each of the newly formed atom of elements.

Then next statement is "and let the dry land appear and it was so done". This is possibly the initiation of the universe that we know today. The term "appear", or "appareate" in Latin is unique in its meaning. In Hebrew it does not say "appear", instead it indicates that it shall be seen. The text could have said let dry land rise but it didn't. The text specifically indicates the idea of an appearance. And that is exactly what would occur if the elements were assembled and gathered on this day. You can use our air as an example. Evaporated water is invisible to us for the most part, and when it is compressed and gathered tightly into a large enough mass we then can see it in the form of a cloud or water droplets.

The same is true of all of the elements. They are not visible until they are gathered into large enough clusters, then things appear. We can witness this in various laboratory experiments as chemical chain reactions occur. The importance of the particular usage of the word "appear", indicating detectable items, cannot be understated. We cannot expect that the accounts of these initial events would begin to explain the finer details of the inner workings of the elements since most of us misinterpret the limited information already given in Genesis One. After all, most people never even read the full creation account which is only several hundred *words*, or roughly only five minutes of reading, so to expect that detailed descriptions should have been given to us, or understood by us, is not rational for us to imagine when this Creation account was given to man from God.

The plants now have a home in which they can begin to develop in the "the dry land earth" which is embraced through the attractive nature of gravity. When considering the order of events listed in the creation account in Genesis One, we have to understand that if this account was is invented by humans thousands of years ago, then either someone had a very good grasp on astrophysics, or they were given information that they likely did not understand. In fact, some of the information that we can deduce from the Genesis creation account is only now possible for most of us to grasp due to our modern scientific equipment.

Every Plant for Itself

The next part of the day three text is another couple of statements loaded with information. The text reads: "he said let the earth bring forth the green herb and such as may seed and the fruit tree yielding fruit after its kind which may have seed in itself upon the earth and it was so done and the earth brought forth the green herb and such as yieldeth seed according to its kind and the tree that beareth fruit having seed each one according to its kind and god saw that it was good". We already discussed that it was the "Earth" that brought forth the plants. Now we want to focus on something that is sometimes debated in the godless-evolutionary circles of science. Did the first amino acids used in the DNA sequences begin from one location on earth where everything came from that single source or one single cell, that against all odds formed?

This part is a somewhat big topic. The odds of godless undirected evolution creating even one unguided living organism are quite low and highly unlikely. But mathematically, given enough time anything is possible... mathematically that is! Is it reasonable to imagine this? In a word, No. It is not reasonable to imagine that a single source or even several sources were the initial points of the origins of all living organisms. As described earlier, life wants to live and it does so voraciously with or without us humans.

When "he said let the earth bring forth the green herb and such as may seed and the fruit tree yielding fruit" the command was given that this should occur. It was not instructed to occur here or occur there, but rather was a general command that likely could occurred everywhere, including any other planets that might have had a suitable environment that would allow the bringing forth of plant life. The statement says "let the earth bring forth". This is a very open statement that allows for a great amount of diversity in location.

We tend to think of this in a local manner, in that the Earth we all live on is a very specific location. But it is unlikely that the "earth" referred to at this point is our own Earth alone, but rather it was most likely referring to any "dry land" that was made to "appear". This would for sure include all bodies in space that were capable of sustaining plant life. This again brings us back to our little mental boxes that we constantly attempt to package everything into. We must open our boxes and peer into the unknown with open eyes and open heart. The "Earth" specifically named on day three was the "dry land". There is no indication that it was only our own Earth. But if we use this in a broader sense then the "dry land" can account for any planet that has a solidified dirt surface or a surface consisting of other matter.

The plants likely could grow only on "dry land" that had "seas" in near proximity. What we miss in this text is that we assume that the "seas" are our oceans and lakes only. But it does not speak specifically of surface water at the point when these waters were gathered and named as "seas". Thus, the "seas" can include subterranean water bodies similar to those we tap into when drilling deep wells for our drinking water. Most people are surprised at the idea that there is a lot more water in our Earth than meets the eye.

In the flood account it speaks of the subterranean waters bursting forth. So, what I am proposing here is no stretch in a Biblical sense. The "Earth" and the "Seas" issue is another matter of scope having to do with our tiny mental boxes. If the water called

"Seas" also included the subterranean waters, then the plants would have had ample moisture to begin growing wherever water was available. But this growth only includes the rooting growth part of the plants and not the mature development of them. The mature development aspect is the part that is filled with wild godless-evolutionary speculation in science, or it is ascribed to "God-did-it" in religious circles.

What doesn't fit in the scientific approach are the odds of the plants occurring at all, and additionally, the odds of similar kind plants occurring all around the world. While it might not be stated so, we can infer from a godless-evolution standpoint that the evolution of specific kinds in more than one place is stretching the theory far beyond reality. However, if we allow the guidance through an instruction, then it is not only logical but also certain that specific kinds would appear in more than one place in the same general timeframe of events listed in the Genesis One text.

The godless-evolution approach is an every-plant-for-itself approach that would have been unlikely to sustain life for long. But since we see how voraciously life wants to live everywhere the conditions warrant, we can easily picture that plants were not in any sort of competition for resources, but rather were brought forth everywhere and enhanced each other all in relatively short order. When the command was given for the "earth" to "bring forth", it is very likely that this was not a single location here on our earth, but was more likely occurring all over our earth; and further, not on our earth alone but on any earth-like body in the heavens where life could propagate from the "let the earth bring forth" command.

Plants Within Plants

The command "let the earth bring forth the green herb and such as may seed and the fruit tree yielding fruit" is an interesting command that allows for the instruction for plants. And as discussed in an earlier chapter, the terms "green" and "tree" are likely far more specific terms than intended in the original text. This means that the plants are divided into two or three categories, "herb", "and such (or vegitation)", and "fruit tree" (or wood or maybe fibrous- stick as mentioned when discussing raspberries earlier). In fact, the important part in this separation event is the division between "fruit" versus other plants.

The next part of the day three events is very interesting "the fruit tree yielding fruit after its kind which may have seed in itself upon the earth". Notice that the fruit has "seed" "in itself". Now keeping to our DNA approach, we can look at the fruits we see today and see that those fruits have seed in them and that seed carries the DNA required for the plants to grow "each after their own kind". This is rather than the seeds growing as a part of the plant leaves or other not edible parts.

The next part "upon the earth" requires us to zoom out of the text slightly and look at the whole thought "and the fruit tree yielding fruit after its kind which may have seed in itself upon the earth". The question is, does "upon the earth" pertain only to the fruit itself? Or is it talking about the fruit plants overall? And, is it possible that it is also referring to the herb and vegetation as well?

If we look at our plants today, we find plants that grow completely underground and also those that sprout above ground. But fruit is uniquely an above ground plant. With the exception of the peanut, all fruit grows above ground, but the peanut being considered a "fruit" is a human-made designation and does not necessarily make it proper designation. With fruit, we eat the fruit which is the bulk of the food rather than the seed, but with peanuts the seed is the main part. So, the "fruit" designation for a peanut is stretching definitions a bit. You might want to infer from this that only fruit grows above ground and that the text indicates that everything else is grown underground. But that is not what I am indicating, nor is it what the text indicates.

The distinction of the fruit plants being "upon the earth" confirms what we witness today, and aside of the misclassification of the peanut, it is very accurate. The seeds that each plant contains have the DNA within them to propagate their kind. They are plants within plants. And those that bear real fruit are always upon the earth, but the herb and other vegetation is not specifically located "upon". Plants, like potatoes and carrots, grow beneath the surface and are within earth, but grass grows upon the earth and grass is not included in the fruit plant category. So as we witness every day, the broad classifications and their locations are constrained to the locations listed in the creation account. Which is that the "herb and such" have no designation and are free to be within or upon earth, but the fruit plants are to be "upon" only, and the seeds are within the fruit part or a part of it.

Chapter 10

Plant Life is Committed

As you have probably noticed, the text of the Genesis One creation account can be taken as a far more detailed account when we read the words as the words that they are, rather than assigning generalized concepts to the overall text. When we dispose of our little mental boxes of preconceptions and read any Bible version of antiquity (meaning prior to the 1700s) we find that you cannot be considered a rational person if you accept replacing the term "firmament" with terms like "expanse", "canopy", "vault", or especially "sky" or "horizon". The text, when read and ordered or arranged as it is presented in these older versions, is very specific, and it is only when we commit to reading it as stated that we can truly begin to understand its many intricacies.

What Are We to Believe

What are we to believe about the Bible's creation account? Are we to take it word-for-word as an actual account of the entirety of creation using our understanding of today's word definitions? Or is it merely a glossed over explanation of what might have happened? Or is it maybe all fabricated by someone who was insightful and made it all up—after the fact?

The last question "is it maybe all fabricated by someone who was insightful and made it all up, after the fact?" is highly unlikely since the order and detail within is really quite scientific. And if not for our orbiting telescopes we would not be able to fully connect the content of Genesis One with what we see in space.

The notable part here is that when we keep our scope of interpretation broad and keep in mind the chicken-versus-egg aspects of the naming done in Genesis One, we can clearly see that "Earth" is named *after* something else was previously referred to as "earth", and that early "earth" substance in the first sentence is not our *Earth* that we live on today. We do not have to pervert the text or change words to come to this conclusion; rather, we only need to understand that this is the likely sequence of events that are described in the Genesis creation account.

This chicken-versus-egg issue in our human understanding is perhaps the single biggest hurdle that we must overcome to understand the text as it is intended to be understood. The chicken-versus-egg issue is questioning: was it the chicken that came first or did the egg come first, which is really a wonderful and very loaded question that is specifically discussed in the *The Science Of God Volume 3 - Day Five and Day Six - The Creatures - Revolution or Evolution*. What we need to extract from the chicken-versus-egg question is not the answer as to which came first, but rather the overall general concept of *cause* and *effect*.

One of the first things that we need to decide when analyzing the Genesis One creation text is to decide if we will allow things done on days three and four to be referenced in advance on days one and two. This is a critical point as it pertains to *cause* and *effect*. This is true overall throughout the entire topic, but in this case, I am referring exclusively to the *naming* that is done during the first four days, especially the third day.

If we believe that our Earth was created in the first sentence then we have nullified all of the remaining text and perverted it. Sadly, many of us do this because we lack detailed analytical skills. This lack of deep analytical skill is not a problem for most people, but it is common amongst us, however, in science we expect hair splitting analysis of everything, yet when it comes to the Genesis One creation account, we fail at that terribly so in science. What are we to think when the "experts", who are "scientists" with doctorates, or the theologians and other Bible experts, can't even separate the *cause* and *effect* issues of the naming conventions in the Bible's Creation account?

God's Handy Work

Who gets to name things? Who gets to translate words? In our world, the person studying who is the first to coin a term is the person who had the "naming rights". And it is the person with enough knowledge on a topic and enough mastery of a language who gets to translate written work from its native language to another language.

When terms are coined or a translation is made, the best we can do is our best to convey the essence of the word or thought behind it. In recent years, in our arrogance, we often set aside this practical approach as we opt for obscure words to name things that lack a descriptive nature, or we use words that are so far from general understanding that they are meaningless to nearly every one without the naming purpose being specifically explained to us. We also have the interference of our lust for recognition by using our own name to give name to our discoveries. But what of the Genesis account?

We have already established that for someone to invent Genesis at the time it was recorded with its specific events order which is in full agreement with our modern scientific observations, would be extremely unlikely. And so we find ourselves left with the likelihood that this text was somehow handed to mankind and written down, and then passed from generation to generation. Given that we are discussing the Bible's Creation account, we have to entertain the idea of a discerning Creator, who in some manner passed this information to us. If this is true, then apparently this Creator wanted us to know this specific information–and *that* is a very big statement! This information was communicated to us and it is very obvious that it is a very specific accounting of the order of events leading to the creation of all things. This is not true of any other creation account found from any other ancient cultures, which incidentally have much similarity to the Biblical creation account, though many have a somewhat childish spin on the account.

The question we are left with in the translation process is, what did the original documents actually say? And further, what did that sound like? God's handiwork is detailed in Genesis One but what we don't know is what was truly conveyed. All we have to work with are the various attempts at conveying that information to each language through translations.

In the western world, we have the Bible translated into many languages, however, the versions of antiquity are few. Greek, Aramaic, Hebrew, Latin, English, and German are the main languages that we can look back on and review in effort to understand the original words and their intent. But having these versions of antiquity does not mean that the words used are the original words or original sounds that were uttered, penned, or scribed. However, when we review these older versions, we can see that terms like "earth" might not have been in the original text. For instance:

Douay English Gensis 1:1 "in the beginning god created heaven and earth."

Latin Gensis 1:1 "In principio creavit Deus caelum et terram."

As you can see, in the Latin translation the term "terram" is used, and the translation of "terram" can be *ground* or *earth*. But nearly every modern translation uses the term "earth". The Masoretic text uses a word pronounced "eretz" and German uses "erde". So, you can see that this issue of translation is not as easy as one might hope. The best we can do is the best we can do, and seeking these older Bible versions and using them in unison is our best approach. It ends up being somewhat like an investigation or presentation in court using collected information to ascertain what is true, what is likely true, and finally what is obviously false or incorrect. We must use the process of deduction of these texts along with our scientific observations to try to determine the potential original words or intent in the text.

The modern presentation of the Latin that I have been using in this book is derived from the Jerome Vulgate, which is the only one that does not use a translated word that somewhat resembles "earth" in its sound and order of letter types. The Vulgate "et" in Latin translates to "and", the Vulgate versions available that are of antiquity use "ettarram" without a separation between "et" and "tarrum" and is typically translated as "and earth".

Unless we can access the original writing, we are confined to using our best judgement and logic to reason through these translation problems as we discover "God's handiwork".

A Commitment from God

The creation account as stated in Genesis One is a type of commitment from God. Not necessarily a commitment to us specifically, but rather a commitment to the Creator by the Creator. It is a set of commands that caused order to abide in the heavens. This commitment is thorough and constant and can be depended upon. We depend on these instructions every day of our lives and we cannot separate ourselves from them while in human form.

We use these commitments and instructions with every step we take where we don't float out into space, and with every breath we take of our atmosphere that is held in place by none other than the gravity that was made apparent and indirectly indicated at least as early as day three. What we don't know is if this commitment is eternal or if it will end. An important point to keep in mind is that if the forces of physics mentioned earlier were to cease, then so would every tangible thing we see. If you really ponder this for a while it is quite a profound thought that everything is held in place by unexplainable attractive forces that if they ceased would cause everything to immediately dissipate.

After Their Own Kind

The instructions in Genesis One are very basic instructions: be light made, separate the waters, gather the waters, bring forth etc. But while simple in the initial reading as we gloss over the text, "bring forth" is very complex when we dig into it. And the instructions further state that, the plants which are brought forth are to do so with "seed after their own kind". The DNA instructions at this point are already laid down, but to what extent? This is where the line between evolution and creation begin to blur.

The plants producing seed "after its kind" can be taken in two ways. The first way is that the seed being referred to is the plants' DNA, and the second way is that the plants produce seeds as we know them today. We already discussed this in an earlier chapter and determined that the seeds carry the DNA regardless of whether it is in the form of raw DNA or in the form of an encapsulated seed that safely protects and holds the DNA. We also discussed that the DNA is in the plant and also within its roots, which can be repeatably proven by taking root material and re-planting and nurturing it as is commonly done in horticulture today.

The most interesting result is that the newly re-planted DNA from the plant will abide by the instructions in the DNA, regardless of where the DNA is from as long as it is good DNA. From a godless-evolutionary standpoint this is somewhat contradictory. Why is it contradictory? Because, while the environment of a plant will affect some aspects of it, such as the flavor of grapes planted in varying soils, the grapes first and foremost are still of the "kind" stated as "fruit" with seed in itself, and second, the particular variation of grape will also remain consistent. But given enough pressure from environment sources, subsequent generations can eventually produce different varieties. All of which is consistent with the day three events in verses eleven and twelve.

What we are told about day three in the Genesis One text is consistent with science on *all* accounts. The arrival of plants on day three matches perfectly with what we see in the fossil record where there are general types of plants, or kinds. Nowhere in the fossil record are these general kinds breached.

Chapter 11

Open and Receptive

Besides us humans opening our minds and being receptive to information that does not fit into our little mental boxes, the function of *reception* is also a part of the creation account. While the "Earth" did bring forth, it did so without *seeds* as we know them today. God did not plant the plants using seeds, they were brought forth by the "Earth" and they "yieldeth seed according to its kind". While the plants would have rooted a great many times over, they were constrained to the few plant types listed and they likely did so all over the "dry land". The plants could not have done much growing above the ground as we see them today until the Sun shone on day four.

On day four "god said let there be lights made in the firmament of heaven to divide the day and the night and let them be for signs and for seasons and for days and years to shine in the firmament of heaven and to give light upon the earth and it was so done and god made two great lights a greater light to rule the day and a lesser light to rule the night and the stars and he set them in the firmament of heaven to shine upon the earth and to rule the day and the night and to divide the light and the darkness and god saw that it was good and the evening and morning were the fourth day".

Day four is more of an astrophysics day, but it is biologically critical. Had the Sun shone before the plants developed, it is possible that the plants would have been unable to actually develop. We have to understand in this regard, that when "god said let there be lights made in the firmament of heaven to divide the day and the night" it is likely that those celestial bodies were already in the process of forming or had already formed when the waters below were being gathered as a part of letting "the dry land appear".

We have our pop-science speculations that the stars produced the more complex elements through fusion and fission, but this could be entirely backwards. If the various atomic elements were all created or assembled on day three, then it is possible that the stars and our Sun were initially part of the "dry land" appearing. And due to the magnitude of size, the pressures of the elements could have begun to luminate through fusion on day four.

Due to the spectral data that we get from observing our Sun and the stars we can speculate on their composition, but since these bodies are far from us, very large, and very, very hot, we will likely never know for sure what they are internally composed of. We don't even know for sure what is deep inside of our own planet Earth. The point here is that there is every possibility that the stars are now and always have been made of many elements, and that those elements become simpler on the surface from fission rather than more complex through fusion. And *that* is an important distinction to take note of. Stars are typically very large and the greatest pressures are likely inside of the star, yet we do not know if the gases are also burning within them or if they are solid material within. We can speculate, but anyone insisting otherwise is still making little more than an educated guess based upon preconceptions and other scientific speculations.

If we peer out of our mental boxes that hold us all hostage, we can then see that all heavenly bodies could have begun their formation on day three. We can also see and that the "dry land" may have been the substantial collection of elements into a solidified state, with other elements forming compounds that remained recognizable liquid and were gathered and called "seas". Remember that earlier it was mentioned that the term "waters" likely had the intent of meaning fluidity.

The statement, "let there be lights made in the firmament of heaven" carries with it a broad encompassing definition that allows for the use of the "dry land" in the making of those lights. Based upon our scientific understanding of physics and astrophysics, this fits very clean into science. Now, we cannot force this to mean that *all* "dry land" *must* have become lights. But rather "dry land" *could* become lights when the "be light made" command was initiated.

"let there be lights made in the firmament of heaven" would be the igniting of the elements composing these bodies. From as far as we have been able to observe, stars are typically considerably larger than planets, and we believe that it is the intense pressure caused by the massive gravitational forces that causes them to heat up.

What we have to question here is, did these "luminaries" do this as a result of the massive pressures, or were they suddenly all ignited similar to the way we start a fire with a match? Here again we have the issue of time to grapple with. Is day four limited to our twenty-four-hour-day since the lights are now in the firmament? Not likely.

Days could not have been our twenty-four-hour-days until the day four events had been *completed*. This is because we don't know how long these lights took to ignite. And if they actually ignited, we don't know if it occurred on every single star at the same time or if they all ignited at their own time and pace. From what we observe in space today, this is an ongoing process that likely occurred in large quantity during the day four events, but may actually have never stopped occurring and might be continuing to this day. If the actions that occurred on days three and four occurred everywhere, then we can assume that they would largely have completed early on, and any lingering matter that is not yet gathered might be rare today but still visible when we look for it.

The development of the "lights" on day four could also have begun occurring as the day three events occurred. I have been using the term "day", but while the "days" certainly indicate a sequence of events, they lack a true distinction of specific amounts of elapsed time. All of the events could have begun in an instant on the given "day" and occurred as the commands were given, but may have taken a very long time to be fully reacted over an amount of time that is far greater than our twenty-four-hour days.

The *being* or coming to existence of light could have occurred in an instant as well as the firmament separating when separating "the waters that were under the firmament from those that were above the firmament". But things are a bit different on day three's "dry land appear" event, because our earthly experience shows that plants need dirt to grow in. But is this true? Do plants require dirt to grow or at least to begin to root? The obvious answer is, no, they do not. Many plants can begin to grow in water alone as long as the water has some of the chemistry that the plants need to begin growing.

So where does this leave us with the overall topic? Am I changing direction, thus nullifying the previous thoughts about land appearing and the earth bringing forth? Potentially, but the bringing forth subject allows for a great deal of variance as to the state of the "dry land". Our experience tells us that plants do not thrive in a dry environment, but "dry" is a relative term. What is "dry"? Some versions of modern Bibles say "God called the dry ground land", but this seems to be a bit off, since the versions of antiquity point to the "dry land" being called "earth". Yet in all accounts it indicates *dry* or *arid*.

"dry" and "arid" are relative terms that infer low moisture content, or lacking fluidity. A slurry of mud could be considered dry when compared to only water, because it in fact is drier than water. We have to consider that dry could be specifically referring to *reduced* of amounts H^{2}O at this point in the text, or even to something simply somewhat solidified. There is a great deal of room for legitimate variation of scientific explanation here that could be a combination or a specific detailed occurrence. But either way, the plants were brought forth by the "Earth" in the text.

The plants being immersed in more of a slurry at their initiation seems more likely because that sort of environment would better allow the assembly of the DNA instruction for the various kinds. Plants being in a slurry of chemistry would pertain only to the initialization of the plant DNA instructions, and an eventual further settling of the dry ground into a state more familiar to us would then allow those instructions to be carried out more fully. Yet, we again run into the problem of needing light and an ability for the plants to somehow receive that light.

Opening to See the Light

On the fourth day "god said let there be lights made in the firmament of heaven to divide the day and the night". In this statement it appears that the lights had specific purpose. The first purpose listed is "to divide the day and the night" and the second purpose is to "let them be for signs and for seasons and for days and years to shine in the firmament of heaven" and the third is "to give light upon the earth".

It is in the day four text that we can be reasonably sure that this refers to the stars or lights that we see today in the night sky along with our Sun. We can know this because it fits very clean with how we measure time today using "for seasons and for days and years".

Perhaps the most important part of this part of the text regarding plants is the statement "to give light upon the earth". We must keep our scope of view very wide here and realize that this is likely referring to all "earth" in the entire heavens and not just our planet Earth. Now, "to give light upon the earth" is done by all of the stars, not just those that we can see from our Earth, but our star, the Sun, is special *to us* in this regard. The proximity of our star to us gives it a unique ability to give our Earth light and heat, which is near to zero from other stars to us. Yet those other stars give light and heat to the many planets that surround them.

The advent of light is the key to the plants being able to mature and become as we experience them today. Prior to the light, the plants could have been rooting and re-rooting in the soil creating a soil that is evermore rich adding the additional nutrients that did not already exist in the soil prior to that activity. Those plants would have likely attempted to creep through the "dry land" surface, but would have not been able to flourish, at least not based upon our current experience and with our understanding of how plants grow. If the above ground growth did not receive any light then the ground might have been covered with webs of plant life of a root-like manner.

Once the newly commanded lights had illuminated and the light actually reached the surface of earth, things would have changed rapidly. As nearly every adult has witnessed and experienced, plants in a well-nourished soil will grow rapidly when they get the right amount of light in combination with their environment. Plants such as corn can grow several inches in a twenty-four-hour day and bamboo can grow a couple of feet per day.

Up until this point in the creation account, the plants could not see light and would have been hindered from full maturity as we know them today, and thus the seeds as we know them today likely did not come on the scene until a minimum of days and weeks and months after the lights were made. This is the point where photosynthesis would have been able to begin. What we don't know here is whether the plant immediately grew into full forms similar to what we see today, or if they had to adapt over long periods to achieve the ability to photosynthesize.

The fossil record is more inclined to be interpreted as a sudden arrival of fully formed plants, but there are problems with the fossil record not mentioned here that are discussed in a later volume of *The Science Of God Volume 5 – Boats, Floods, and Noah - The Deluge*. Regardless, the fossil record indicates a sudden arrival of many plant forms in a fully developed form, rather than through some long-term slow progressive intermediate process. When light reached "earth" throughout the heavens, plant life flourished quickly in an environment that in many places was likely at a point that would have been chemically nutrient rich and ready for life to thrive. Once the plants' leaves grew, a whole new world of mature plants was born.

Receiving Pollen

There is a property that pretty much everything discussed so far shares, which we have not yet discussed directly, and it is that of reception or receiving. When a command is given it must be received. Most everything mentioned in the Genesis account of creation *received*, including the Creator. The Creator received the indication that it was good through seeing it. Observance is the act of receiving.

Reception is a part of the nature of creation, the only place that we do not have specific indication of reception is in the first sentence of Genesis One "in the beginning god created heaven and earth and the earth was void and empty". But everything following indicates some sort of reception, either through God's observance or through receiving commands, and those commands were followed by the aspects described, and even nothingness received the "heaven and earth". In understanding this undeniable aspect of nearly every living thing in existence, we have to also realize that to *receive* requires an action of giving or offering, we cannot escape this basic Truth. God offered commands and those commands were received. Everything follows this pattern, and very noticeably so–plants receive as a primary activity.

Many plants produce pollen which is a sort of DNA carrying seed, that without, the plants won't produce the fruits in their fullness and will not produce the seed that encases the DNA instructions that is used to grow plants from the seed "after their kind".

Ears of Corn

"Ears" of corn is in interesting term. In reference to "ears" or hearing or taking in and receiving, the hair on the end of the developing corn-cob-sprouts, receives the pollen from the corn tassels. Corn is a type of grain and its kernels are the seed that carry the DNA for the plant to be reproduced in large quantities from scratch.

Simply placing a kernel in the dirt will, within about five to ten days, produce a new sprout of a corn plant that will grow and produce more kernels which can in turn each produce its own new corn plant. Each corn stalk can produce several ears of corn and each ear of corn can produce roughly eight hundred kernels. So, from a single seed of corn we can produce thousands of corn seeds. The actual seed count will depend upon the variety of corn and will vary from cob to cob and will vary with the environment that the corn is grown in. This amazing reproductive system that is instructed to produce seed each after its own kind, ensures that life will live on and corn kind will continue and likely cannot ever be completely eradicated even if every human tried to do so.

Once the attempt to subdue a particular plant ceases, it is almost certain that within short order we would find the particular type plant popping up everywhere corn fields have ever been planted. Not in the quantities that we see in the cornfields of today, but certainly we would see some corn plants sprouting up. And if just one of those plants were to grow to maturity and its kernel-seed somehow scattered, it could produce hundreds of corn plants, all within about a year's time. Now consider the second generation of corn from that single plant; the second generation could be about a million fully grown corn plants in only two years, and then, over a billion corn plants the third year.

Corn, or grains, are very generic and fall into the "herb and such" category as the seeds are not "within" to be considered in the "fruit" classification, such as is the case with an apple. It is the seed itself that we eat when we eat corn and other grains. The same thing is true of other grain plants where they rapidly reproduce and can, within only a few generations, be producing billions of plants from the original single starting seed.

So far in life, having grown up in a rural community, I have seen no discernable change in any of the various crop types that are commonly planted in the field annually. This consistency adheres to the guidelines laid down on day three of Genesis One "having seed each one according to its kind".

Why Are Most Plants Green?

On day three the plant sprouts would likely have been white if they made it out of the ground at all, but we would not have been able to see them without light, thus they would have had no color at all since there was no light to be seen or to reflect off of them. It is only after light with a spectrum of some sort appeared that *color* can be detected. Prior to light, everything would have been what we think of as black.

However, once the lights with a rainbow of spectrum became visible light and that light reached earth, then colors could be physically seen and produced. Now just because there is no light does not mean that what we think of as color would not have been present if suddenly the light was visible. We have to consider carrots or radishes that grow underground and yet are quite vivid in color. But again, while those plants do grow underground, their above ground leaf systems do serve a larger purpose with regard to the capturing of the solar energy and the photosynthesis of the overall plant.

Photosynthesis is the key to our environment as we know it today. Much of the richness of colors in nature can be attributed to photosynthesis and the Sun, combined with the environment that each plant has grown in, the amount of light it has, and finally the DNA with which it is built.

Using water, sunlight, and any nearby chemical available, plants process these substances, and through those substances the plants grow and grow and grow. During this process, plants cleverly reconfigure the compounds as they strip off some atoms and replace or combine them with others to create new chemical compounds, such as oxygen and sugars. The photosynthesis process allows for the plants to form various types of chlorophyll, which in combination these various types of chlorophyll offer plants their diverse spectrum of greens that we see in all areas of nature.

The processes that occur during photosynthesis and those that also occur as a result of photosynthesis, are too vast to discuss in this book and are better fit for topics specifically aimed at botany or microbiology.

Not only do these processes and the chemical compounds that they produce offer a very wide variety of shades of green, but they are also responsible for creating the chemistry that produces the vast full spectrum of colors that we find in plants both above and below the ground. For instance, the orange of a carrot, or the orange from the fruit that we call "orange", or the sometimes-orange flowers that we decorate our homes with. There is no end to the varieties of plants that offer us the beautifully rich spectrum of colors that we see each day in plants.

We humans ourselves have proven the unlimited diversity with which each kind of plant can produce through our deliberate environmental changes and pollination modifications. We do this work at our workbenches in our laboratories, and we have proven beyond doubt that there is little limit to the color differences that can be produced through deliberate circumstantial changes and cross-pollination. We further must realize that many such changes can also occur naturally, such as when the Sun becomes more active or when temperature fluctuates and stays cooler or warmer for a longer period of years.

There are also environmental effects such as a seed being eaten and subsequently trapped in bird droppings, as well as the bird possibly having taken a seed from a warmer climate and then planted it through their bird droppings into a region with cooler climate area. This is not exclusive to birds, any animal can carry seeds in this way, and that animal's body chemistry will in some small way affect that seed and the seed can and will be carried to a different location than from where it was eaten, sometimes very far away.

All of what was just mentioned is only a very brief glimpse of the countless ways *kinds* can vary to produce the richness that we see today. And based upon the fossil record, they have produced diverse varieties early on, which is a topic in itself.

Chapter 12

Calibration

Calibration is a form of indexing to organize to a specified set of rules. All phases of the Biblical creation process included some form of calibration at a most fundamental level. Calibration becomes our guide for everything. Everything in our physical world works off of a calibrated set of instructions. If this was not true then nothing could be relied upon. The calibration in actual physics and astrophysics are the master-set from which all other calibrations descend during the creation process. The first known concept of calibration was the realization of its need as stated in Genesis One verse two when "the spirit of god moved over the waters". Verse two says "the earth was void and empty and darkness was upon the face of the deep and the spirit of god moved over the waters". This can, without forcing the text, easily be understood as an assessment process of the previously produced "heaven and earth" and a determination to begin a type of calibration process that set the standards for all things thereafter.

The first calibration was the allowance for light to "be" as in "be light made". It is a point of definition utilizing the "waters" of the "heaven and earth" material established in verse one. The second point of calibration is a division of light and dark when "he divided the light from the darkness and he called the light day and the darkness night". Regardless of the specific words that might have been used in the original Genesis text, if the light was divided in this way and distinguished as "day" and "night" then it was, in fact, calibrated to the specifications to which it then abode.

Calibration is everywhere and in everything, and it began at the beginning of creation. Calibration is a refining process rather than an instant set of given parameters, meaning that it progresses with each subsequent calibration or recalibration event.

Astronomical Measurement

From our perspective here on our Earth, we are familiar with days and years. Our calibration for this is one complete point-to-point trip around the Sun on our Earth, which defines what we recognize as a "year". Next, we use our Earth's spin on its axis as another calibration that we call a "day", which is one full zenith to zenith revolution or a "twenty-four-hour day". These calibrations were likely laid down on days three and four in the creation account in Genesis, however, there is really no indication of that level of specificity, and to insist that it could only have been during those two days of events, limits the text where there is no limit stated or needed. Yet, it is somewhat logical to propose that that was the case.

It is important to note that while our pop-science view insists the stars somehow gave birth to the planets, there is no conclusive proof of this. And the planets, according to the Genesis account, occur either before or along with the formation of the stars, including our Sun.

On day three when the "earth brought forth", our Sun was likely not yet giving any light but may have been largely gathered and possibly was close to its current size, but not yet ignited. Now I realize that this could be scoffed at by many astrophysicists, but it is those very same astrophysicists that claim that the big bang is a cyclical event, and further that since everything was compressed into a ball of matter so small that it essentially did not exist, yet it still was able to bang. Thus, when it was in its smallest form it was not hot, and could not have been since no movement could have occurred given its theoretical size. The point here is that it is not a requirement that the stars be lighted only due to their mass and the pressures exerted on the elements of which they are comprised. If we insist that heat is absolute when under pressure then that would mathematically force the big bang to have been very hot and glowing when in its singularity state. But this is an anomaly in the theory since the laws of physics supposedly did not yet exist in that theory.

The stars could very well have been at the ready and then could have been forced into ignition due to massive impacts from other bodies, thus initiating atomic chain reactions of fusion and or fission that continue to this day.

The astronomical time Measurements we use today were put in place on day four when "god said let there be lights made in the firmament of heaven to divide the day and the night <u>and let them be for signs and for seasons and for days and years</u> to shine in the firmament of heaven and to give light upon the earth".

What we miss in this particular event is that it is likely where we go off the rails in our understanding of the Bible's creation account. This particular text gets a bit tricky here because the particular calibration of "day" matches the vocal or audible terminology that is used on day one when light was created, and so we improperly connect these two very likely disconnected uses of the term "day". We must further hold fast to the understanding that this particular part of the Genesis creation account likely applies to *any* solar system *throughout the entirety of the heavens*. Other planets in our own solar system have zenith-to-zenith day lengths that are very different than what we experience here on our own Earth. The issue of calibrating the term "day" here is critical in being able to properly understand the creation text, and it is again a chicken-versus-egg cause-and-effect type issue.

On day one, the light was divided and named along with the darkness, but it is only on day four that this calibrated division was potentially first used in a human visual manner. With light being created on day one, it was likely the light *function* that was created, but on day three there was nothing yet to light. The calibration instituted by the division of light is critical to understand. Since light as we know it today is a part of the electromagnetic spectrum that we are familiar with, the "divided the light from the darkness" may possibly have been the institution of this spectrum or rather the calibration of it.

Instituting a calibration rule does not mean that the particular calibration will have an immediate effect unless the calibration is able to immediately fully affect that which it is intended to control. But while light was "divided" through the calibration process, we simply do not know the fullest extent of this division.

Now, it is critical to understand that the electromagnetic spectrum that we know today was not necessarily the day one division calibration specifically, but rather the idea behind the electromagnetic spectrum is likely similar in its nature. The electromagnetic spectrum calibration is there whether or not it is in use at any point in time. Further we must realize that when we speak of light and how it works, it is only when the light is in its photonic state that it is "visible" to us, but as it travels via its "wave" it is *not* visible, so this is another potential explanation of what that particular day one division accomplished. In fact, it is possible that without this particular division, the electromagnetic spectrum could not exist because it is what we named "waves" that make up that spectrum.

When translating, we are limited to the Bible versions of antiquity that are available to us today. We are also limited to the phrases, words, and tongue to which they were translated into. So, when terms like "light" are used in the text we have to question what "light" is, which we can partly do by reviewing Greek, Hebrew, and Latin. From an English language standpoint, our best option is to first go to the Latin because it uses a character set similar to English, and many of the words are recognizable to us. Latin uses the terms "lux, lucem, luminaria, luminare, illuminent" all of which translate to "light" or "lights" in English in the various places that those terms are used in the creation account. This simplification presents us with a bit of a problem during the interpretation process.

Two Plus Two Will Never Equal Five

The terms "lux, lucem, luminaria, luminare, illuminent" all have a different purpose in the Latin text. "lux" is only used in verse three and is referring the light overall before any division. The next usage is in verses four and five which uses the term "lucem" when it was divided. The next use is "luminaria" in verse fourteen when "god said let there be lights made in the firmament". And then in verse fifteen in describing the purpose of the "light" it uses "illuminent".

Why does two plus two equal four and not five? To put it more clearly, why does 2 + 2 = 4 and not 5. The reason this is so is not because of the arbitrary Arabic numerals used. Those modern-day numerals have nothing to do with the values of the outcome of two and two is four. Whether or not we personally are conscious of this, in our minds, words describe concepts and those concepts are represented by the words we use to describe those concepts. For some concepts, the concept is embodied within the word such as the word sounding like the sound that it attempts to describe such as "thump". Other concepts need multiple words to convey the underlying concept to us in English.

In our English-language-minds we use the term *light* or *lights*, such as when we say street "lights" to refer to the lights that reside alongside of the city streets. But we also say that something "lights" our way. One usage is a concrete item and the other usage is the function of that item. But we also have the result of that function which is the actual light that has been cast.

A numeral is an abstract representation of II plus II resulting in IIII. The symbol we call "two" represented as "2" and the symbol called "four" represented as "4" are nothing more than symbols to represent physical items or the representation of those items or other abstractions in our minds. The same is true of the various concepts that are represented in the Genesis creation account wherever the English term "light" is used.

The original text was likely first written in an ancient Hebrew text or something that may have predated that. Any subsequent translations are just that—translations, and they have lost some of the original descriptors for the concepts surrounding the usage of the term "light" in our modern English Bible versions. But just as two and two can equal any symbol we choose to make them represent, the concept behind two plus two will always be II plus II equals IIII. We can call IIII whatever we want, but the value will always be IIII. So, while we can change terms such as "lux, lucem, luminaria, luminare, illuminent" to "light" or "lights" during translation, the underlying concepts behind them will always remain.

The problem we have when altering the words used to represent a concept is that some of the original words may have contained additional information that we could use if only we understood the original word used. This leaves us trying to fill in the conceptual blanks as we flounder in our understanding of the Genesis One creation text.

This same thing applies to everything in the creation account, and if you think back to everything written in this book so far, the majority of it pertains to this *underlying-concepts* issue. But do not let this discourage you in any way. The underlying concepts that are in any translation of the Genesis creation accounts in Bibles of antiquity are all reasonably consistent, thus, demonstrating to us that the information has been carried forward as well as can be expected when translated.

The advent of light did not do the action of actually luminating something until it occurs in verse fifteen when it was made to "shine in the firmament of heaven and to give light upon the earth". It is only then that the plants could receive their needed light to begin to substantially grow up robustly through the ground and begin their photosynthesis process.

Mile High Review

If you recall, in an earlier chapter we discussed mile high stack of paper, with each sheet representing one thousand years in the proposed age of the universe of roughly thirteen to sixteen billion years old—an age that was proposed by science when this book was written but is likely to change as people awaken. Our own Earth's paper stack is claimed to be roughly only one quarter of that entire timespan which is a stack 1320 feet tall. But in truth, we really do not know the true age of our Earth, and any age proposed is only scientific speculation.

Earth's age estimate is based only upon radiometric dating of material that we have access to and our speculations of the duration of time that the deposited rock layers took to form. But the Earth could be far younger if the radiometric calibration is not correct for long-term age-analysis. Further, the radiometric dating can only be done on superficial material that we have access to, and since the earth has a diametric radius of about four-thousand miles and we humans have only drilled down to record depth of just less than five miles into the Earth's surface, we simply cannot say how old our earth is.

Earth could be far older than we speculate. But there is another factor to consider which is: When exactly does the counting begin in the age of the Earth? Is it the age that the elements that Earth is composed of were made? Or is it when they were gathered? If it was when they were gathered, then at what stage of the gathering would it begin to be considered an entity of its own? Was it when the elements were still drawing together and not yet in the form of a solid ball? Or was it after it coagulated enough to make something that was more solid? Was it when the material was hot from activity? Or was it when it had cooled enough for plants to begin? **When** do we being the counting?

As you can see by the questions just put forth, it is an absurdity that we will ever know the age of our earth. We might be able to find signs of the very first plant life and radiometrically date them, but that will always be an unknown because we will likely not ever know for sure if there is something older that we have not yet discovered. And we must also realize that our radiometric dating methods leave a great deal to be desired.

One thing that we can be reasonably certain of is that all of the celestial bodies that occupy the heavens are all likely very old, perhaps billions or even trillions of years. We simply do not know and it is very possible that knowing is not possible.

I realize that this might offend some people, but it is possible that not even God knows the actual age of the Universe that we live in today because "lights made in the firmament of heaven to divide the day and the night and let them be for signs and for seasons and for days and years". Was this done for us? Or maybe since there are entire galaxies that we cannot see with the naked eye and only know about because of our recent powerful space telescopes, we can assume that it was not for us, but rather was for the Creator to be able to understand and record duration of existence in cycles. Even if we use our most powerful telescope, there still are likely galaxies larger than our own that we cannot see because they are too far away from us. This does not then appear to be for **our** purpose alone that we can use them to "be for signs and for seasons and for days and years".

For our human purposes, dating plant remnants is our best method to guess at the age of our planet, but using radiometric dating to gauge the age of earth using plants and rocks has its limits.

Layers of Deceit

It is truly perplexing to listen to some of the pop-scientists discuss the layers found during excavations in archeological digs. Each layer is typically attributed to many years of sediment, often thousands or even millions of years of sediment per layer depending upon the individual scientist assessing the layers. These numbers are absurd. Layers are laid down quickly and most of the plant fossils that we find are found *between* the layers, where one layer was laid down and then a plant grew or a leaf fell onto the surface and was then covered by a subsequent sediment layer.

The view of each layer taking more than a few years each is ridiculous in the highest degree. While we cannot know the time between sediment layers, we can, just from simple logic, determine that the layer itself was laid down rapidly. This is because while plants and small animals trapped between layers evidently were on top of an existing layer, they are also trapped within the subsequent layer that lay on top of them. This occurs if their physical structure was strong enough to overcome the weight of the newly deposit layer that came down upon them, ultimately encapsulating them *within* that layer, rather than only being sandwiched *between* the two layers. Further, impressions between layers are generally impressed on both the upper **and** lower side of most items found between layers.

While many layers are quite thin, only a matter of a few inches, some layers are very thick, and yet the same holds true– items entrapped are either between layers or typically are encapsulated within the layer that buried them if their physical structure could withstand the weight of the material they were covered with. Further, since some structures pierce multiple layers it is additional testimony that the layers had been rapidly deposited. We know from simple logic that the layers were laid down in the order we see them with the oldest layers lower in the stack and the newer layers nearer to the top, this of course excludes any of the outliers or over turned layers mentioned in an earlier chapter.

Common sense and pure logic demand that we know that each layer was deposited relatively quickly, likely in weeks or less, rather than in thousands or millions of years. Yet, we still have the unknown time gap between layers, but even that is cut short due to the fact that above and below layers have impressions and there are many instances of items piercing multiple layers of sediment, items that would have decayed in only a matter of months in some cases and decades in other cases. But no layer in itself shows tens- or hundreds-of-thousands of years for it to have been fully deposited.

Offensive Science

Deposition time for layers is certainly not millions of years each. We can know this from our short earthly human experiences. If you lived in a rural community for more than a few decades, you will likely have noticed the erosion and will have witnessed the layers deposited in low-lying areas of fields as you saw some of them slowly fill in a bit.

I will not debate with pop-science that these layers were laid down as we see them or that the layers themselves might be very old, possibly millions of years old. But that does not make each layer thousands or millions of years in the making. The layers could have all occurred in only a few thousand years total time, but may have been laid down in that length of time millions of years ago.

The pop-science timeframes often promoted to those who do not study these things are typically outright lies and/or misinformation. The deceitful numbers given are all built upon the fragile house of cards known as the "geological column". It is not so much the geological column itself that is offensive, but rather it is the calibration thereof. Radiometric dating is partially built upon the early estimations of the age of the progressive layers within the geological column. But scientifically, we believe that we can tell these ages by radiometric dating methods that use half-life counting of elements found within an encapsulated item.

However, everything estimated is done with a perspective from our current-day understanding of such half-lives of atoms. There is a certain amount of atomic decay that occurs and that takes a certain amount of time with each element. The remaining number of atoms in items found encapsulated in rock is measured and compared with what we might find in a similar plant today. And using the difference between the item found in rock and what exists today, we can calculate the estimated time of encapsulation by comparing it to the known half-life of the particular element counted.

When trying to understand science topics, it is important to ask rational questions regarding our ability to objectively look at the information that arises due to our inquiries. We need not specifically agree when discussing or debating, nor do we need to talk the other person into seeing things the way we each see things. What is important is that we can each objectively analyze the information. Our conclusion is irrelevant to the information and items found and/or to our ability to look at all information objectively. However, when it comes to our conclusion, that's when the problems between people occur and the offensive science or offensive religion begins.

Whether you are a creationist or a godless-evolutionist, at some point you will likely come to a point of *belief* in what you think you have seen in the presented evidence. Your interpretation of that evidence is only a belief and may or may not be true or accurate. The debates and fights come in when one believer attempts to force his or her opinion on another person or will not allow that other person a voice. And this is where the troubles become very personal. For instance, if you have a godless-evolutionary belief system and you try to force your beliefs and teachings on other people's children and those people do not believe in godless-evolution, you are then attacking their children and could lead those children astray if your analysis is incorrect. Similarly, the reverse is true, if you believe in godless-evolution and your children are indoctrinated with 24-hour-day Creation, the parents of the children have reason to be offended because it runs counter to their beliefs whether true or not.

But, neither position has any bearing on what the truth actually is. It's not important as to what we conclude, the important question is, have we asked the proper questions? And maybe even more important is, have we asked the questions properly? Our personal conclusions will be based upon our questions and our openness.

Let Logic Be Your Guide

As you think about the creation account as laid down in Genesis One, let true logic be your guide. But as you do so, you have to be sensitive to the not-so-obvious details that are included in the text.

The details included in the sometimes-peculiar language of Genesis One's creation account are generally common through most versions of the Bible up until the many new versions began to appear during the 1800s. In the 1800s when printing became very inexpensive and much higher speed, we as a people began to toss aside this *history* for *intended* clarity through the adjusting of the text. Various translators adjusted the text to fit each translator's own perception as they chose to understand the text while translating from Old English to contemporary English. These translators' distorted understanding of creation, has, sadly, now been written into the hearts and minds of millions of unsuspecting Christians, mostly evangelical Christians. Many of these Christians go on to get challenged on their six-twenty-four-hour-day beliefs only to be utterly crushed by "science". This causes them to subsequently depart from God due to these inaccurate Bible versions and the inaccurate beliefs instilled through those Bibles.

Our departure from common sense has been going on for a very long time. And at least for the past couple of hundred years more and more people have bought into the foolish idea that creation occurred either in six twenty-four-hour days or that it all happened in a big bang. One view says that God did it in six of *our* twenty-four-hour-days, and the other says that it occurred without God over very long periods of time in a big bang, but both are wrong and both are right.

It obviously did not occur in an instant with a big bang long ago, and it certainly did not happen in six of our twenty-four-hour days, but every single aspect of it does speak of design and order through following explicit instructions on a broad scale, which demands intelligence to carry out.

Knowing what we know and have clearly proven about the speed of light today, it is certain that the universe is old and that it took time for the light that we are seeing today to reach us. So logically, we know that the visible universe is very old when describing it in our Earth years. And when reading authoritative versions of Genesis One through an unbiased scientific lens, we readily see that the creation account as stated in Genesis One makes several pure logic statements that fit with observable science today, as well as points of logic that some things must come first in order for the next event to occur.

Use your pure logic, or maybe better put *logic-of-purity* to inquire about the creation account. And remember that everything is calibrated to follow patterns set forth during the events listed in Genesis One's creation account.

Chapter 13

So Many Plants

We already discussed the "waters", but let's dig a bit deeper into that subject. As indicated in an earlier chapter "waters" is almost certainly a *property* of the "heaven and earth" in the very beginning of Genesis's "in the beginning god created heaven and earth and the earth was void and empty and darkness was upon the face of the deep and the spirit of god moved over the waters". Terms like "deep" or as some Bibles state it, "bottomless", implies direction. But Latin uses "abyssi" or our "abyss". These subtle differences lead us to imagine deep water, but if the text intended *watery* or *fluidity* similar in nature to a fog being everywhere, and *abyss* instead of "water" and "deep" as we understand those terms today, then we could understand it as the extents of space which is the more likely concept being conveyed in that text. Our problem is that while something can legitimately be translated to "water" and "deep", it does not mean that the translator actually intended for us to receive it as we use those terms today or, more importantly, how the translator might have misinterpreted what the text actually intended.

Fluidity is a *property* of water. Water has molecules that are able to flow when grouped in large quantity. An important part of the fluidity property is *quantity*. Anything that does not chemically crosslink and bond will flow, provided that it is not frozen and that there is a large enough quantity. The fluidity property can be partly that of quantity. The fluidity property is a fundamental property shared by all living things in their inner workings.

The Flow of Life

The flow of plant life occurs only due to fluidity. Life flows from one generation of plant to the next as the DNA is passed from cell to cell from within each cell, which is also true in creatures. As the water is absorbed by the plants and carried throughout the plant using capillary action, it allows for the exchanging of atoms in the compounds that plant is both utilizing and creating. It is the free flow of these materials that causes the plants to offer their vibrant colors, diverse shapes, and their seeds and succulent fruits.

Without this free-flowing property nothing could live. The property of fluidity is what allows grass and the branches to sway in the wind stretching their fibers during the flexing movement. Without the ability for things to flow, elasticity could not exist and grass would snap and break clean off the moment it moved in the wind.

The chemical compounds that would have gathered in the statement, "god also said let the waters that are under the heaven be gathered together into one place and let the dry land appear", were available to the plants brought forth in the statement, "let the earth bring forth the green herb and such as may seed and the fruit tree yielding fruit after its kind". When we limit this particular text to the "waters that are under the heaven be gathered together" as being H^2O alone like in our seas, there is no other place in the text for other chemical compounds to form.

We might, in our small-box-minds, want to force <u>all</u> H^2O into the "gathering together of the waters he called seas" statement, leaving absolutely no trace of H^2O in the "let the dry land appear" after the "seas" where named. But, as our current day experience clearly shows, this is not the case and it likely never was since the gathering began occurring when "the waters that are under the heaven be gathered together into one place" actually took place. Once the command for the "earth" to "bring forth" was instituted, each "kind" likely immediately began assembling itself or began to be assembled by the *earth* which was the result of "the waters that are under the heaven be gathered together".

The fluidity property includes the flow of light, the flow of energy, the flow of water, the flow of heat, etc. As the light shone upon the earth its heat would have flowed from cell to cell and its various wavelengths would have initiated many molecular processes that transpire during the above-ground growth of all above-ground plants. This would include any plants that have any of their parts that sprout above ground. As these plants grew and ran through their cycle of life they would have continuously fertilized the earth more and more with each plant generation. This would also have been true of the pre-sunlight plant rooting that would have taken place for a long period of time between the "let the earth bring forth" statement and the "to give light upon the earth" statement. It is the *fluidity* property that allowed for all of this to occur.

A Sea of Grass

Our interpretation of our word "water" can be an action, such as watering the lawn, or it can be the property of fluidity as in something is watery, or it can be the actual water or H^2O itself. The term "seas", in our modern view, is quite specific in its meaning. To us a "sea" is a large body of water. In Latin the "seas" are referred to as "Maria"

Douay English Gensis 1:10 "and god called the dry land earth and the gathering together of the waters he called <u>seas</u> and god saw that it was good"

Latin Gensis 1:10 "Et vocavit Deus aridam Terram, congregationesque aquarum appellavit <u>Maria</u>. Et vidit Deus quod esset bonum."

In Hebrew "seas" is translated to sound like "mayim". In Greek there is "hydor, hydro, nero". The point here is that we must use our logic and common sense to ascertain the underlying intent in the text using multiple languages and multiple Bible versions along with *commonsense-scientific-analysis*. It is the particular nuances in the words used in the various versions and languages that can clue us into the original concepts intended to be brought forward in each word usage. Let's take for a moment the Latin "Maria". Since it is a name that was given, is it then proper that it should be changed to "seas" when translated to English? But further, what was the Latin translated from and what would have that sounded like to our ear in its native language? There is even deeper digging that can be done when interpreting any words Biblical or otherwise, but for brevity we only lightly touched on that subject which could fill many volumes.

Water or watery is an issue of movement or fluidity like the waves we see as we drive by a lake or the ocean. We also see such movement directed by the winds as we drive by grain fields or open prairies on a windy day, as in "amber waves of grain". Fluidity is everywhere, it is life, and it is what gives us amber waves of grain and seas of grass and ocean waves.

Who Will Care for the Plants

Do plants need care from humans? To some extent they do, but plants will live whether or not we humans are here. The plants existed long before humans did, likely thousands or millions of years before. Do we know for sure that plants came before humans? Yes and no. Biblically, the order of events is clear and places humans as the last point in the order of creation. But, scientifically we also have our geology. When we examine the layers, we see evidence in the fossil record showing that the plants and animals seem to appear before humans in the lower layers and humans finally appear later in layers nearer to the top.

That geological evidence is very compelling for the order of events and the arrival of humans. However, there are flaws in the geological information depending upon how we calibrate the geological record and how we read that record. This is discussed in the later volume, *The Science of God Volume 5 - Boats, Floods, and Noah - The Deluge* having to do with the Genesis flood. Yet it seems that no matter how we want to view the order of events, whether Biblical or godless-evolution, that us humans came *after* the plants and animals. So, since us humans came after the plants, we have to ask ourselves: Could it have been possible the we would have been able to live on a barren planet void of all plants and animals?

To get a grasp on that idea, just imagine trying to live on our Moon, or planet Mars. The moon has no atmosphere, but Mars does have an atmosphere, although it would not support us breathing as we do. But regardless, Mars is currently without any plants as far as we know, and without those plants we would promptly meet our end if trying to live on Mars without packing lunch. This alone is indicative of the plants having come before humans. So, what's taking care of who here?

Yet, while plants take care of us humans, it is also true that we humans can care for the plants, and when we do, the plants flourish in a way that they generally would not in the wild. We have been given dominion over the plants and thus we can manipulate them for our own benefit. We can make plants thrive by how and where we plant them and by what we feed them after they have been planted. We can group plants for cross-pollination and arrange them, especially flowers, to beautify the landscape.

Plants are wonderful devices brought forth by the earth that started on day three and they have not ceased since. It is our place to care for these plants and enhance them to benefit the earth and mankind.

Where Did the Water Come From?

The water in "and the spirit of god moved over the waters" is almost certain to not be the water that we splash and play in today. As mentioned earlier, the languages used in various Bible versions use terms like "seas, mayim, hydor, hydro, Maria, nero" when discussing the water or aspects of it. In Latin, Genesis one verse two uses "aquas" for "water" and in verse ten "aquarum" for "water" and in Latin it was named "Maria", and "seas" in English. It is the "Maria" that are the "seas" that we splash and play in today.

Notice the two different words used in Latin verses two and ten, "aquas" and "aquarum" respectively.

Douay English Gensis 1:2 "and the earth was void and empty and darkness was upon the face of the deep and the spirit of god moved over the waters"

And

Latin Gensis 1:2 "Terra autem erat inanis et vacua, et tenebrae erant super faciem abyssi : et spiritus Dei ferebatur super aquas."

Douay English Gensis 1:10 "and god called the dry land earth and the gathering together of the waters he called seas and god saw that it was good"

Latin Gensis 1:10 "Et vocavit Deus aridam Terram, congregationesque aquarum appellavit Maria. Et vidit Deus quod esset bonum."

Yet in verse six "aquis":

Douay English Gensis 1:6 "god said let there be a firmament made amidst the waters and let it divide the waters from the waters"

Latin Gensis 1:6 "Dixit quoque Deus: Fiat firmamentum in medio aquarum : et dividat aquas ab aquis."

The English translations, while very well done, do lack the nuances of the languages that they have been translated from. In all places where the term water is used in most authoritative English versions, "water" is used in the plural form "waters". But in

Latin we have "aquas, aquarum, aquis, aquae" which in English all translate to "waters". Often in languages there are nuances defining gender and plural that are unlike English. But there are also other nuances that simply do not translate using single words in English. This is why we must seek to understand the underlying meanings and we must not rely only upon a single Bible version as our source when studying the creation account.

Logically, the "waters" gathered as "seas" could not have been H^2O until, at the soonest, when the gathering of "waters" was initiated. Interestingly, at that point, the term "aquarum" is used in Latin which is like our term "aquarium", which is a pooling of water or a body of water. In our human nature, we always assume that we need massive forces to bond atoms to create molecular compounds, yet this is done on a constant basis, silently and efficiently, through chemical reactions that occur in many natural processes day and night every day, especially in plants. So, while we can force things to occur with our sometimes gruff human methods, nature does so with seemingly little effort, yet it is unlikely that the plants and other natural processes we see today initially made the water.

The atoms that comprise water, *hydrogen and oxygen together,* are very explosive when mixed as separate atoms and ignited. Yet when bonded together in sets of two hydrogen atoms bound to one oxygen atom, they then become the water that we love to drink and play in. The bonding of these atoms likely occurred during the gathering of the waters in "god also said let the waters that are under the heaven be gathered together into one place". Stepping away from the mental picture of "dry land" rising out of "seas", consider that day three may have had several stages of gathering in the overall gathering.

Douay English Gensis 1:9 "god also said let the <u>waters</u> that are under the heaven be <u>gathered</u> together into one place and let the dry land appear and it was so done."

Latin Gensis 1:9 "Dixit vero Deus : <u>Congregentur aquae,</u> quae sub caelo sunt, in locum unum : et appareat arida. Et factum est ita."

We can easily break this gathering into at least two groups. The second would be the gathering of the actual H^2O as it settled into the "seas", but the first could easily be interpreted as the gathering of the many atoms into specific groups, with each group being one molecule of H^2O. But there could also be a gathering stage before the first mentioned here, which can be attributed to the day three events as well, and that is the gathering of the subatomic particles to form what we now call atoms. Then at that point various specific solid objects, or atoms, are quite possibly a form of the "dry land" referenced in verse nine where "god also said let the waters that are under the heaven be gathered together into one place and let the dry land appear and it was so done". Verse nine is an event that was completed with "and it was so done". Then the subsequent naming occurred in verse ten.

Water or H^2O as we know it today was, in a real scientific sense, most likely "gathered" on day three, first by the various atoms being assembled, and then subsequently by the grouping of these atoms into sets of two parts hydrogen to one part oxygen. All of most of the other atomic elements would likely also have formed during that period, thus allowing the subsequent formation of a diversity of many other chemical compounds that make up the world that we live in today. The final gathering would specifically have been a gravitational gathering due to now fully formed atoms, allowing for the pooling of H^2O into lower areas causing the formation of "seas" or "Maria".

It is possible that there were no open pools of water on the surface such as we today would refer to as oceans or lakes or seas. *Subterranean* gathering of H^2O was possibly the primary gathering of waters that existed at that time. Also, the land was potentially all somewhat saturated with H^2O within the surface allowing for the bringing forth of plants by the earth. As we know today, plants will not thrive or even grow in truly dry land that is absent of all H^2O. The amount of H^2O in the so-called "dry land" is up for speculation, but given that plants can live in water alone, we can safely speculate that the "dry land" has a meaning somewhat different than we what we understand today, because as mentioned earlier, "dry" is a relative term. This view allows the plants to flourish everywhere on the face of the earth in abundance, thus allowing for the required flow for plant life to quickly begin as the earth was commanded to "bring forth".

Chapter 14

A Plant's Life

According to the Genesis creation account, the plants were brought forth before there was any light to shine upon them. This was discussed earlier, where plants do not need light to create root systems. Plants only require H^2O, a bit of warmth, some nutrients, plus instructions to do what they do. Then once the lights were made "to give light upon the earth", the plants could begin their above-ground activity to their fullest extent.

Plant Activity

What is life? And what are its parameters? In the laboratory we seek to create life in our, thus far, futile attempts to make a single living cell able to replicate itself or even be considered alive. But while our floundering attempts are intriguing research, nature does this trillions upon trillions of times every second of every day all around the world.

As we seek to understand the origins of all things, we try to define an ever-sharper line of what "life" actually is. Here we're not referring to this on a human growth basis where we question when "life" begins, as it does "in the womb", but rather we are specifically referring to plant life here and the cells that dictate their outcome.

"Life" as best as we can understand it, is an animate function. In other words, it replicates and to some extent *moves* through that replication. This is true on a fully-grown plant level, but in this section we are referring to the cellular level of the plants. In science we have subatomic particles, atoms, molecules, and finally cells in the broad picture. It is the activity of these cells that makes something "alive", and without this cellular activity the chemical compounds would be similar to solidified rocks in that they do nothing animated.

When you peer into a cell and see the inner workings, you see each part seemingly blindly doing its designated activity, likely as instructed by the DNA. But much like when breaking down atoms, when you break down cells into their component parts each of those parts is not living in the way we think of something living. If you were to dismantle a cell it would no longer function and would just be clusters of parts made of chemicals. This is much like if you were to dismantle a running automobile engine piece by piece. Once it is stripped down to its component parts, it can no longer function.

It is the active function of each part, when properly assembled, that allows an engine to run, and the same is true of *all* cells. It is the combination of parts, fuel, and DNA that allows for cells to function and live. The engine of a car is only functional when it is fully assembled, and it is the ignition system and fuel that allows it to run. The ignition system in an engine along with the camshaft and crankshaft timing *instruct* the engine in order to make it run without further human intervention.

Living Plants

When something is alive it typically has a grand intended function that constitutes the whole of life. An engine will do as instructed, but the engine is limited to that instruction only and it will eventually wear out and, in essence, die. Cells, on the other hand, have a unique ability to do two things: The first and most important is that they can replicate. And second is that they can self-repair in most cases when they have access to the proper nutrients. And those are both actions that a car's engine cannot do. It is the ability to self-replicate and self-repair that truly makes the living cell "alive". The DNA instructions direct the cells' movements through the structure's system, and DNA instructs their ability to replicate, their placement, and more. When a plant is alive, it is an entity unto itself and can grow and mature. It feeds from its overall environment and is somewhat influenced by that environment to create a diversity of its "kind".

Plant Offspring

As the plant grows and matures, it packages its DNA instructions into safe little packets of potential life. Just add water and new little replicas of the plant take root into the nearest source of nutrition. When in a natural habitat, that nutrition is typically found in "the dry land earth". The plant's offspring will repeat all of the processes that the parent plant carried out until it reaches maturity, when it will produce its own safe little packets of potential life and cast them on the ground for the "earth" in order to allow the next generation to be brought forth. This process will repeat for an unknowable amount of cycles for time infinitum.

This is what we see today. It is what we expect tomorrow. And it is what we see in the fossil record of yesterday. All kinds of plants will do this as they replicate using the forces of environment and intermixing of DNA to create vast varieties of each "kind", offering each generation its own unique DNA instructions, giving us unlimited beauty to enjoy. As each plant completes its full instruction given through its DNA, it will meet its end, usually after it has shared its DNA for the subsequent generations.

Death of a Plant

The death of a plant is an important part of the life-cycle of plants. As the plant goes through its instruction, it absorbs water and nutrients and when it protrudes above ground the sunlight that is cast upon it assists in building cell upon cell until it has developed a root structure that will tap deeper into the ground to draw in more moisture and nutrients from the soil. As it grows, it consumes the nutrients and builds itself a strong structure as prescribed by the genetic instruction in the DNA. And it will do so until the DNA instructions have been completed, at which time the plant will forfeit all of its nutrients as it wilts and falls to the ground to be broken down by cells of bacteria that dwell within the ground and will consume it to become even more nutrients for future generations of plants. This is indeed a very brilliant system!

Chapter 15

Self-Replication

Self-replication is the key to life, and without self-replication the "life" of anything will promptly end in a single generation of that lifeform. Some things can potentially self-replicate on a cyclical basis without self-replicating cells in their makeup. But this sort of self-replication is not done by utilizing a DNA sequence to instruct it. Such non-cellular replication is done in laboratories through reciprocal activity that balances chain reactions that can continue to oscillate due to those chain reactions. Cellular activity, on the other hand, that follows DNA instruction is the key to actual living organisms and their ability to replicate as far as we, to this point in time, have discovered. It is the DNA sequence instruction sets that allow "kinds" and life to propagate.

Funding the Research

Our curiosity is the driving force behind science... mostly. Once upon a time, we humans sought to truly understand the Universe in which we live and the planet on which we live, but somewhere along the line we have lost our way. When reviewing history, we find that losing our way is common in history and it will continue as long as arrogance exists.

There was a time that people who did science did so out of pure curiosity and devotion to the Creator. There was also a time when the Church funded science. But as is common, money and arrogance tend to foul a pure idea. There are those who propose their erred theories that they have spent the greater part of their adult lives building and promoting. In our modern era, these people receive a great amount of funding from government grants and private endowments etc. Often this funding will also self-replicate as long as the errors in the research can remain alive in the hearts and minds of the people. They deceive both the giver and receiver of these funds, and all of the people involved have their reputation at stake when errors in their theories are proven and made public.

If those offering funding find out that they have been supporting flawed and even outright incorrect theories, then they will lose a great deal of respect from the general public who support their campaigns or their companies. Additionally, those who receive the funds are likely to lose their "professor" status and tenure at colleges or their positions at organizations that do such research. They will also lose their public pop-science presence and influence that they so crave. This career crushing problem only occurs due to our arrogance, and until we stop the self-replication of our lies, we will remain trapped in this mess and be unable to find the truths that we truly seek.

And the Like

To advance in our understanding of the heavens and all things seen, we must self-replicate that which advances our understanding of Truth. When we fail to replicate seeking and finding truths, we are then certain to have our failed theories eventually die and then we must live with the humiliation of our public errors.

It really does not matter what we are referring to, the issue of self-replication is perhaps the single most critical aspect of everything we do. All of creation, aside of the calibration, is ultimately instruction. But even calibration is instruction within instruction.

When something true is able to be replicated, it leads to flourishing, and improvement, and life, and understanding, and truth. But when the replication is false, it is sure to quickly fail, and as long as the facade can be perpetuated due to the blindness of the general populous accepting those errors, the lies and errors are able to continue. Yet, as always, eventually the lie will be exposed in a way that cannot be denied. Then those who perpetuated the lie will be forced to admit their errors, or they will look evermore foolish for denying the obvious truth.

The Bible's Genesis one makes many clear statements that we write a great deal of nonsense into in our assessment of that text. And more than anything, it is our inability to accurately translate the text that has caused so much debate regarding creation. Phrases like "and such" are very broad and allow for a great deal of variance. We can suspect that while the creation account is very broad, it is actually quite specific, though not detailed. So, vague statements like "and such" were not likely the original intent. The original if present, was likely somewhat more specific and intended to cover any plant life not included in fruit and herbs.

The Firmament

To not get too far off topic, the "firmament" is that which becomes a foundation. We do see this firmament in the structure of science with *money* being science's firmament, but when it comes to understanding creation, the term "firmament" is more important than just about anything else in the text.

"Firmament" is a unique term that doesn't really have any single translation word, nor does it need translation. To make firm is to make stable or maybe to make solid. When "god said let there be a firmament made amidst the waters and let it divide the waters from the waters", the "firmament" somehow caused a distinction between two types of "waters", those above and those below. Latin uses the terms "super" and "sub" rather than "above" and "below". Instead of thinking of this in terms of the directional designations of "above" and "below", think of it in terms of *superior to* and *submissive to*, respectively. In other words, it is possible that if it is *superior to* the "firmament" and/or the lower/sub "waters" it is not subject to it, then the "firmament" and/or the lower/sub "waters" have no control or power over it, but it if is *submissive to* then the "firmament" will guide it. The *superior* waters are simply *superior*, and the *submissive* waters are *submissive* only because some of the waters are *superior* to the *submissive*, meaning that some of the waters are somewhat more pure or somehow different and higher quality than the other waters.

This is fairly obvious when we open our mental boxes and peer outside to see the truths laid down before us. The "firmament" being a form of calibration sets a pattern that controls most things. Things become firm, or ordered, when they are subject to a given instruction. For instance, when plants' cells replicate according to the DNA within them, they will increase and follow the specified pattern set forth in the DNA until that pattern is fully carried out as the reading of the DNA sequence is completed. Once the Sequence is completed then the firmament of that plant has ceased, and the plant will begin to decay until it has completely dissolved and returned itself to the earth.

The Evidence Supports It

As we observe the pattern of patterns in all things, we see that the evidence supports the idea that the "firmament" is a type of fundamental calibration that divides types of "waters". This makes perfect sense and is entirely logical in the physical and also explains the potential undetectable "Heaven" where many believe angels reside.

The Bible speaks of the "father above" or good things being from "above". Scientifically, this should at minimum be respected, though it would, in this case, not be detectable with any scientific equipment due to the fact that it is superior to or "above" the "firmament" or maybe above the lesser "waters". It is completely understandable that science has little interest in that which cannot be detected *scientifically*. Physical sciences are just that, "physical" sciences and they deal only with the physical, and thus the "waters above" do not fall within the framework of "physical" sciences. Physical things can be touched or seen by us.

From a Biblical standpoint, all evidence supports the concept that the firmament is a function or power that divided the "waters" by calibrating the *submissive* waters by coalescing them into the matter that we are scientifically aware of today, and it has possibly done so without affecting the *superior* waters in the same way. The Bible has many references to "Heaven" and "God" in conjunction with the directional term "above". We might want to rethink our concept of prayer being directed to something up in the sky, and instead direct our prayer to something superior to that which we are familiar with in our fleshly tangible world. "above" is not a location, it is a state or quality of being.

The entirety of the Genesis creation account supports the idea that the "firmament" is some sort of calibration, that without, nothing that we can sense with our five senses or our scientific equipment could exist. Once we open our eyes to this basic view about the "firmament", it is difficult to undo our understanding. Once truth is known, it is difficult to un-know it.

Even though the "firmament" might not have directly affected the *superior* "waters", it did indirectly affect those "waters" by distinguishing the *superior* from the *submissive* "waters" or the "waters below". In a sort of heaven-driven rationale, we can consider the *superior* "waters" as a sort of material of perfection that no contamination can touch. Not necessarily as an obscure command from some megalomaniacal "God" that insists on perfection, but rather as a result of the "firmament" that was the cause of the desired divisional calibration. This could mean that it is likely not possible for any worldly impurities to enter into it because it is continually separated out by the "firmament".

From a Biblical perspective, it is clear that the "firmament" is something special that "divided" the "waters" and allowed the "waters under the firmament" to begin their process that largely continued their completion processes through the time-duration included in the days three and four events.

Chapter 16

Perspective of Plant Evolution

Our human perspective is key in our understanding of anything we encounter. For instance, this is true even in our emotions: If a child is reared in a home of violence, they will have a difficult time understanding love because their perspective is that of cruelty. And the reverse is true: A child reared in a loving family will not fully grasp the cruelty experienced in a home built upon cruelty.

If you have read *Hot Water – Your Perceived Identity - The Life Repair Manual* you are likely familiar with the fact that our individual human perspective alone affects *every* facet of our lives. This basic truth will never change for us humans until we ourselves allow ourselves to seek only Truth.

When we have perspectives that we have built up in our hearts as "fact," it causes us to hold prejudice against that which fails our prejudicial tests. It does not matter if you are a Christian-creationist or a godless-evolutionist, you can be exactly correct on your perspective, but still be closed minded.

Our scientific perspectives are no different. If we fail to have an open heart, we will fail to see truth. This is true of both science and religion in regard to the topic of origins and creation. When we discuss evolution, it is usually from a godless perspective, thus not allowing for any possibility of intelligently guided evolution.

Out of the Lips of Scientists

While you may not have had the opportunity to personally interview any pop-scientists, you most likely have read or heard their own words from their own lips through interviews and video programs, or books they wrote that you may have read. The words that they speak from their own mouths are their words and thoughts, even if those thoughts have been derived from previous pop-scientists.

The words that we speak can have long lasting effects on those who listen to those words. If we claim that we speak "fact" and we do so with the illusion of the voice of authority, then our words will affect the rationale and thought processes of the willing receiver who believes those words. We are ultimately responsible and accountable for the lives of those to whom we speak. If there truly is a God, and if our words are false, then we will pay a price for our deceptions to our fellow man. And if our words are of truth then we will be rewarded for sharing that truth with them.

But scientifically determining truth is a bit tricky when we cannot directly witness an event such as our Earth's initial actual bringing forth of the plants, or an imagined big bang. What can we produce as evidence? And is that evidence verifiable?

Claims of Statistically Verifiable Evidence

In pop-science we often hear of claims of statistically quantifiable and verifiable evidence. But, what exactly does that mean? We touched on this subject earlier where predictions are often thought of as "evidence" when we happen to find the things that we predicted we might find. However, the predicted evidence could have been the result of a different cause than what we theorized. We can make statistically quantifiable predictions about astrophysics, but in truth, many of those predictions are not *verifiable*.

Much of the "verifiable evidence" when predicting is wishful thinking when it comes to theoretical astrophysics, this is because it is currently unlikely that any human being from our Earth will ever live long enough to be able to travel far enough to collect the predicted evidence first-hand and then return that evidence to those to whom the prediction has been given. So, while we can *mathematically* make many predictions about how things came to be, we will never be able to fully prove these predictions, thus much "verifiable evidence" in pop-science suffers from the definition-of-terms dilemma. *Who* gets to decide what is satisfactorily "verifiable"?

This is somewhat different with plants than it is with astrophysics, in that we can make certain statistical predictions that we can verify because those plants can be touched and tested, and generally these predictions can be verified in only a matter of weeks or months through growing plants. Yet, when we claim godless-evolution and speak of things millions or billions of years back, we are then limited to the materials found between layers of sediment as our remaining tangible evidence. But the way much of that evidence is evaluated is left wanting.

Choosing Your Evidence

In the science-versus-creation topic we hear about "cogitative bias" which is to accept that which supports you and reject that which does not support you. This is often brought to our attention when referring to creationists, but somehow seems to be forgotten when referring godless-evolutionists.

Each origins discipline must choose what evidence they feel will "prove" their theory. This makes perfect sense, until a real challenge is brought forth that defies our chosen evidence and our concluded theory. If we fail to address this new challenge, then we have deliberately closed our eyes to potential truth and can no longer claim to be objective.

You might think if perfect evidence was presented, that the person in error would immediately see the truth and admit to their error, but this is typically not the case. There have been more than a few instances where people have been clearly caught on camera doing a crime and yet they still deny that they did so, which is often how people approach the science-versus-creation debate.

From a pop-science standpoint, there is no possible evidence that will force us to allow any thought of a God into godless-evolution. And with creationists, there is sometimes no allowance for any sort of evolutionary process whatsoever or any change in their creation model view.

If you are a true-faith-creationist you *know* that *science* and *creation* not only can coexist, but in fact, **must** coexist for creation to be taken seriously and for it to be true. But in doing so you must choose your evidence wisely, rather than cleverly.

In pop-science, there is a deliberate effort to disallow *any* thought of a God, and thus that discipline has chosen to not allow any evidence whatsoever that might suggest a Creative intervention of any kind. Any information or thoughts with an inclination towards a Creator will be altered and re-explained in a way that removes a Creator from the information and from all consideration. This exact same tactic occurs with some creationists when they invent theories that remove the possibility of any type of progressive evolutionary change from the discussion.

"Evidence" is subject to our interpretations for whichever theory we have chosen to promote. But what does the geological record actually say?

Dolostone

Things in our interpretation are slowly changing, sometimes for the better, and sometimes for the worse. The absurd idea that layers took thousands or millions of years to form is fantasy-science. Geologists fall into that trap, but it is those who *don't* study or understand the formation of rock layering and its basic logic and physics who readily accept most of the very long-age estimates of sedimentary rock.

The layers tell a story, and we have perverted that story with our inaccurate assessments of those layers. We find manmade items that are familiar to us today buried in some layers, but we refuse to acknowledge that each layer was rapidly deposited. I am not sure of the reasons this persistent insistence continues today, but there is a great deal of obvious and logical evidence suggesting that sedimentary rock is quickly deposited and solidified.

There are materials like sandstone that are compacted and very easily eroded, which in itself is very telling. If it took a great deal of time between layers, then erosion would be extremely common and very noticeable in those sandstone layers due to how soft they are. But as the layers typically show, the sandstone is generally consistent throughout and does not show much if any evidence of erosion during the alleged time frame that it was covered with subsequent layers that consisted of a more rugged material.

Materials such as dolostone were likely deposited rapidly and solidified through chemical reactions and some heat. When carbonate silt or mud is combined with magnesium in water it causes a cement-like hardening of the silt.

Cement is not an invention of humanity; it is a discovery of nature. And if you have ever poured raw cement powder into a bucket of water, or even just poured the powder onto the ground and let it sit as powder open to the elements of weather, you will quickly see a very familiar look on the surface. Once the rain comes and wets the cement powder it looks strikingly like the surface between layers of limestone. The same is true of the powder poured into a bucket of water, except it will not have any sort of marking from rain drops.

Certain mineral powders, when heated, become sensitive to water and will chemically react causing them to solidify after exposure to moisture. We have perfected this for our human uses to a point where we know how long it will take to cure and become hard, plus we can adjust that cure time through various treatments and combination of materials.

The hard, stone layers that we find in geology today are most likely the result of volcanic activity that cast volcanic debris onto large areas of water and/or were cast out in massive dust and ash clouds. The heated powder material quickly settled into soft layers that were later rained on or wetted in some way and subsequently chemically reacted into a hardened state. There is little denying this obvious chemistry because we have witnessed this in recent volcanic eruptions that we have actually watched occur with our own eyes and recorded with video cameras, and subsequently studied in recent times, such as with Mount St Helens in the state of Washington in 1980.

The Black Layer

It is said that there is a thin black layer deposited that had covered the majority of our Earth's surface at one point. In most areas of the world this layer is said to be found at the same approximate position in the geological record. The black layer is easily spotted and is unique in that it is black versus being in the range of stone colors typically found in the stone layers. But there are other layers that also cover massive regions of the world.

Something that affects an area the size of a continent is suspect of occurring very quickly. When an area is very large, then the potential for major erosion in various parts of the larger area is inevitable over long periods of time. For a layer to stay relatively consistent throughout any large area, such as a major portion of a continent, like the layers we find today, is clear evidence that any subsequent layer was deposited shortly afterwards.

The black layer said to be found worldwide shouts of a very quick subsequent deposition of other material. If a layer of some sort of soot-laden ash was deposited world round, and it stood for more than a few rains there would be very little evidence remaining in many areas, yet the black layer is claimed to be found consistently around the globe. This is clear evidence of the rapid deposition of subsequent layers.

The Earth is a Living Document

This Earth is a living document of the history of the geological events. But just as in our modern age where some people choose to interpret The Constitution of the United States different from what was obviously intended, so too do we interpret the document of the Earth differently than is shown in the obvious recording of layers.

What we suffer from in our modern pop-scientific world is specialism. We see this in the medical field where doctors specialize in a specific area of medicine, but generally only one specialty area per doctor. When you visit that doctor, he or she can only help you with his or her particular area of expertise at which they excel. So, then you are sent to another expert for other problems, and this works well in many cases. However, all too often, the real problems get missed because doctor *A* and doctor *B* are not communicating properly. The result is "misdiagnosis", often resulting in further complications that are sometimes deadly for the patient.

While it generally won't end up causing death when studying geology, the same holds true. People often attend college for specialty work in the various sciences and they are taught certain information as if it is all "fact". But they cannot comprehend things that do not mesh with their specific education's particular discipline. When this is the case, we tend to invent rationalizations to force what we cannot fit into our tightly closed narrow-minded educated mental boxes so that we can avoid admitting our errors and avoid forcing ourselves to forfeit our religion of pop-science.

In medicine, a general practitioner is usually better suited to care for you than a specialist is. The general practitioner will send you to a specialist when required, but they usually have a more personal and better overall understanding of your individual circumstances. This is true in science also, and especially in geology. There are those who specialize in a given area and they do a very good job in that work, but it is the person who has experience in *many* scientific fields who can assemble the various specialty studies into something that actually fits with reality.

Your Scientific "Ah-Ha" Moment

Our "Ah-ha!" moments don't come easy because we insist that we are right and others are wrong. Sometimes it's easy to demonstrate that others are wrong, such as with the big bang and its major flaws. But proving we are right is a bit more complex.

From a Bible believer's perspective, proving that there is a God is often foolishly done by saying the "Bible says so, so it's true!" But this fails scientific tests on several levels. The Bible is not proof of a Creator. The Bible is the record of that Creator's work, but if there is no provable Creator, then the Bible could just as well be a bunch of invented stories. But is it? Is the Bible verifiable?

Mostly yes, the Bible is a fairly detailed account of history, most of which is indisputably accurate to logical minds. If you dispute the histories stated in the Bible, then you are required to also disregard all other intermingled histories of people who have interacted with people mentioned in the Bible. That includes the history of all Jews and Arabic peoples, all of Europe and Russia and northern Africa as well as Asia and many other areas of the world.

The Bible clearly has real significance, but it is the spiritual aspects where it is difficult to prove some things are true. For instance, there is ample evidence that a person who we call Jesus was crucified. But the question is not *if* that occurred, the question is, was he the actual promised Savior? That question only has relevance if the Creator who we refer to as "God" actually exists. Without the existence of this God, nothing in the Bible has much significance, especially the supernatural points mentioned in the Bible.

In current science, our scientific ah-ha moments cannot really ever come because that ah-ha moment will be a clearing of vision and a revelation of the Creator, so in science it would be more of a Biblical ah-ha moment. But for creationists the scientific ah-ha moment comes when we come to realize that many of the modern versions of Genesis One have been altered to a point where the actual meaning cannot be ascertained through certain versions due to inaccurate translation of certain words within those versions.

Until we come to the full realization of the modern translation errors and correct them in our understanding, creationists will continue to be made to look like fools in the eyes of the world, and rightfully so. I realize that this sounds a bit harsh, but if you don't understand creation, then please don't publicly discuss it as if you do; and even more so, do not attempt to teach others, because teaching error tends to eventually turn people away from God and the Bible, thus cheating them out of true understanding. The scientific ah-ha moment is when we realize that our interpretation does not fit real science, but we are still able to see that when the text is properly understood, it may in fact all be true; and while somewhat light on details, it is still highly accurate.

Chapter 17

Spreading the Seeds

The seeds of creation are many and they cover the range of creation, but the seeds of the plants are specific. We discussed the DNA-seed aspect that would have had to have been a precursor to the actual seeds that the plants would produce at maturity. It is irrational to imagine that the Creator formed seed and then planted them like we do today when we plant our gardens. We know through experimentation that very small amounts of material can be taken from a particular plant and used to clone nearly identical plants, so it's a not stretch to realize that plants did not need to originate from the seeds that they produced in their maturity. This is a logical thought, and further, it is completely in line with the Genesis One day three events.

Understanding the initiation of the plants has been a quest for us humans for centuries. Biblically, the earth brought forth at the command of the Creator. The only leap of faith here that we need to take in the entire discussion is, how did the actual DNA of the various "kinds" begin? DNA is an instruction set, which as far as we can tell is a very clever format, and we can attempt to describe it any way we'd like to, but in every description of DNA we use terms indicating intelligence. Now this is not proof in itself, but it bears consideration when we fail to be able to describe DNA without some indication of intelligence and extreme order. The command given in "let the earth bring forth" has no indication of any limiting factor regarding the dispersion of the plants. And from our study of the geological record we see that plants arrived spontaneously <u>all over our Earth</u>, which is consistent with the overall indicators in Genesis One. The command either occurred or it did not, meaning that either there is a Creator or there is not.

Godless-evolution is highly unlikely, but it is not the evolution part where plants can change over time that is unlikely. The unlikely part is the initiation of similar plants *everywhere*. The godless-evolution scientific explanation lacks a great deal of logic. The chain of accidental events that would have to occur makes it scientifically and statistically unlikely for a single plant cell to ever begin due to the odds of those circumstances occurring through pure happenstance. Those odds decrease exponentially with every additional plant location found around the world, yet we must also consider that this same evolutionary initiation process would have had to occurred *identically* in many places around the world nearly simultaneously for the pop-science model to be valid. What reasonable explanation does godless-evolution offer for this worldwide phenomenon? How did similar nothing-to-fully-randomly-**evolved**-DNA encompass the entire globe within the short period indicated by the geological record? It might have been a long time ago, but according to pop-science the record indicates that it all occurred in short order worldwide. Herb kinds, other vegetation kinds, and fruit plant kinds all occur simultaneously in the geological record. This is not something this book is insisting, rather it is what is taught in pop-science culture every day.

Being First to Share

While the godless-evolution explanation for the first steps of life to begin can explain some things, the unlikely very rapid spread of that life is problematic in godless-evolution. However, if experiments attempting to demonstrate the possible start of life via the creation of amino acids is at all accurate, then that could occur everywhere around the world simultaneously. If we can do that in the lab, then nature can surely do it better, quicker, and everywhere. But that still fails us in explaining the similar DNA found around the globe and the various specific kinds that we see when we consider godless-evolution.

Godless unguided from the ground up evolution is not logical in any way. We cannot allow a term that has been commandeered by people who have an agenda to prove that there is no Creator to solely possess and control terms such as "evolution". The idea of progressive change is not new and is implicit in every step of the Creation account in the Bible.

The "sharing" being referred to in this section refers to both the sharing of DNA as well as the sharing of ideas. When DNA is shared, it is typically done through the seeds of plants. But the *initial* plant DNA is a different story. Our science might have the laboratory creation of amino acids somewhat correct, and that activity could in fact have occurred all around the globe in the natural environment. But if evolution is free to do as the environment dictates then we would likely **not** see limited similar DNA "kinds" all around the world. Yes, plants do share their DNA through their seed, however, it is very likely that the DNA did not start from a single location, but rather identical or very similar DNA occurred in many places around the Earth at the same time. And judging from the broad nature of the Genesis creation text, it is likely that the same process is available on any planet throughout the entirety of space that was able to support such growth activity.

Sadly, those who are first-to-market with their proposed theories and who are best at marketing their ideas to the public are typically those who profit the most and are able to best advance their ideas, correct or not. And their inaccuracy in that case is irrelevant to truth. To them, all that matters is that they get to promote their theory and profit from it. I have no problem with this process, unless other people are shouted down and not allowed to share their own competing theories.

Prohibition of Discussion

We have a real problem in our world where free speech is hindered or censored. In America the government is not supposed to inhibit free speech or free expression, yet this is perhaps the most common thing done by the government when it suppresses religious viewpoints by not allowing them to be expressed in government-funded schools. The very corrupt university system and many Ivy League colleges are leaders in such tactics.

We also have voluntary suppression of freedom of scientific speech by the world of media who often will deliberately not air certain opinions and theories. But with increasing sources for media consumption this is not really a problem any longer because there are many venues that will allow such points of view to be expressed. The problem with this is that now we find that there is no longer any vetting of proposed ideas and thus any person with any amount of understanding, or even no understanding, can spread their ideas even if those ideas are highly incorrect. If they present their case well, even if it is wrong, many people will be fooled into believing their errors.

Do Scientific Journals Tell All?

Perhaps the most troubling area of suppression is in the field of science. This includes scientific entertainment that directly suppresses and or attacks any points of view expressing guided-creation, as well as the scientific journals that often do the same. If reasonable theories are being suppressed by any venue, then they are not really telling all, and thus probably should not be trusted. Sometimes there is good reason they don't want to promote someone's theory. For instance, six-twenty-four-hour-day creation stretches the bounds of logic when reading Genesis One and also when studying actual science, so it stands to reason that no one would want to promote it in a science journal.

So, while it is understandable that some theories are altogether rejected, it is not rational to completely disregard guided creation, which includes guided evolution. Guided evolution is evolution that follows parameters, or calibrations, that were set forth in "let the earth bring forth the green herb and such as may seed and the fruit tree yielding fruit after its kind which may have seed in itself upon the earth and it was so done and the earth brought forth the green herb and such as yieldeth seed according to its kind and the tree that beareth fruit having seed each one according to its kind".

Interestingly, the Bible's very ancient account of origins fits perfectly with science's findings, but not with pop-science opinions. Evolution exists, but it is guided, and every bit of scientific evidence we have found thus far declares this to be true. The plants, to this day, abide by the three kinds stated in Genesis One verses eleven and twelve and have not veered from them at any point in the geological record. Those three kinds are "herb", "and such" or *general vegetation*, and "fruit tree" bearing fruit with their seed in the fruit. In its vague description, Genesis One allows for guided evolution to quickly produce many variations of those three kinds of plants that "yieldeth seed according to its kind" and "having seed each one according to its kind".

While Genesis One verses eleven and twelve are vague, allowing only the three generalized "kind" categories, the text also does not restrict variations within the kinds from having been instructed to occur in that account. It is a very open account of the command, and we must look at the key fact that natural seeds produce only after their own kind and we do not see natural deviation from this model. And these "kinds" and variations of them do not spontaneously produce different plants. This is very important in the analysis of plant origins, because many variations of plants suddenly appeared on our Earth as recorded by the geological record, also demonstrates that this is a fairly quick process that followed the guidelines stated in "each one according to its kind".

But this sort of discussion and dissertation of the text seldom appears in any scientific journals, even though it completely matches *all* verifiable science. The journals *do not* tell all.

Explaining Yourself to Get Published

Getting published in journals requires a great deal of explaining using facts and figures and also a great deal of speculation. But if you fail to provide the numbers and formulas then it is unlikely that you will ever get published in those journals. It is not the intention here to discredit them, because they do contain a great deal of true and useful information. But not everything published is true or accurate, because the theory itself, while perfectly formatted and presented, might be wrong; yet if it is perfectly presented then it is likely to get published if it appears plausible and is not Biblical in nature. We can theorize something and we can prove it mathematically and we can prove it through prediction, but that does not necessarily prove it to be actually true, this is because other alternate explanations for a result can be claimed to be the true cause as well.

If you care to get published in a secular scientific journal, you will need to lay down your argument using secular methods to prove your Biblical perspective, and you had better have every thought clearly explainable in your head without saying "The Bible says so, so it is true" or there is no chance of ever being published by the pop-science journals or taken as credible by science or its audience.

You will have to be able to explain yourself intelligibly and you will have to explain how science fits with the Bible's creation account. You cannot expect to get published if you cannot align with *true* scientific findings. To prepare in your mind, you must separate scientific *findings* from scientific *opinions*.

Chapter 18

A Plant's Harvest

When plants finally grew after the earth brought them forth, they had "seed" as each "yieldeth seed according to its kind". As discussed in an earlier chapter, this would have included the DNA *in* the plant *before* the seeds as we know them today had matured, as well as the seeds we are familiar with that do in fact contain plant DNA instruction sets.

But while we can scientifically explain what occurs, we fail to explain why it continues to follow the calibrations set forth. The plants' desire or ability to produce a harvestable seed containing DNA instructions that will produce ever increasing yield is not understood. We might be able to explain it, but we do not truly understand it.

Desire to Know with a Desire to Grow

Increase or *advancement* is a desire that is inherent in all of creation. Consider plants, there is nothing we can do to stop all of them. You can spread poison in your garden and kill everything you planted, but within a few weeks or maybe months new plants will be popping up all over. Those plants might be considered weeds, but they will nonetheless be there and they are vegetation whether or not we like them. We think of "desire" as a human trait where we *desire* to know things. In fact, this entire subject of creation arises only because of our desire to understand. But is it possible that plants have desire to continue? Or maybe that it is the Creator's desire that they continue?

Plants do not need us, but we do need them. And while we could attempt to kill all plants on Earth, if we succeeded in doing so, the human race and all animals would cease to exists in a matter of months. And then the plants, undaunted by our destructive work, would quickly return and fill the Earth. The Earth, while not human and it being without a human soul, still must bring forth as was desired and commanded by the Creator on day three. And this is, in fact, what earth does without fail.

The Joy of Plants

When the earth brought forth the plants it must have been a joyful event. Think about all of our futile efforts to create "life" in the laboratory thus far, and how we cannot seem to even successfully **copy** a single living cell and have it be a *living* cell, even though we have trillions upon trillions of them to examine and copy using every material *already* available to do so. We can clone cells using cells, but we cannot yet copy or produce a single living cell from scratch using designs we currently find in cells.

The Creator must have felt great joy when plants where brought forth. And that joy would have likely been far greater after the lights were made "to give light upon the earth" allowing the plants to mature. We feel proud and have joy if our just-add-water-seeds grow a single flower in our garden, which many of us are unable to accomplish with our not-so-green-thumbs. Can you imagine the feeling of even having imagined the plants and made them be brought forth with all of the plants' diversity, and having nothing to springboard your ideas from?

When we walk through a forest or a well-kept flower garden we often get a feeling of Joy from the beauty and intrigue that we see around us. We see vibrant colors, diverse shapes, and smell sweet fragrances as we pass by the various plants.

There is no reason for plants to be, and it is unlikely that they would exist without the desire, command, and particular calibration of "let the earth bring forth the green herb and such as may seed and the fruit tree yielding fruit after its kind which may have seed in itself upon the earth".

Gathering the Evidence

A part of the living record of the history of our Earth is written in stone. Not in a Ten Commandments sort of way, but rather, the evidence has been gathered for us in the deposited layers of rock. Just like you can deposit money in a bank to be saved for later use, sediment has been deposited and saved for later use as proof. But we fail to use that proof properly.

As we gather our evidence, we arrange that evidence to fit our own desired outcome, rather than accepting the outcome that the evidence shows in its natural order. We force our godless-evolution or our short-term creation into our scientific models and then we take that perversion of science and force it onto the people.

The evidence shows agreement with the day three events. And our gathered evidence tells us that plants have remained relatively consistent from as far back as we can tell in the geological record. Our experiments in botany and horticulture also prove beyond a doubt that the plants evolve within the parameters of the calibration set forth in each "kind" in "according to its kind".

The best part is that most of the evidence grows before our very own eyes and has been doing so for generations. And let us not forget that there is much evidence that has been carefully gathered for us and preserved for thousands of years all around the world.

The DNA of the plants has been gathered for convenience and expedience in the form of the seeds we use today. And it is in these seeds where the grand evidence that supports the statements in Genesis One verses eleven and twelve resides, as each will fall upon the earth, take root, grow, and again bear "seed after its kind". It is irrefutable that this, in fact, does occur each and every day on our Earth.

Reaching for the Stars

It is interesting, the many sayings that we humans have that mirror creation and its effects. When we have a dream or goal that we work to achieve, we often refer to it as "reaching for the stars". We usually feel pretty darn proud if we even get anywhere near our lofty goals.

Even the plants reach for the stars. You might think that this is not so, because the plants tend to draw back or close, such as is seen with flowers when they recede at night when the stars come out. But in this you must realize that our Sun is a star and the plants explode into a vibrant world of color when the Sun rises. And many plants will follow and turn towards that star, our Sun, throughout the day as it traverses across the sky. We also witness a reaching for the stars on a forest's edge and along rivers etc., where trees will have peculiar shaped trunks as they grow in their attempt to harvest sunlight.

There are many "suns" in space and there are likely many plants in space on "*exoplanets*" that are capable of bearing life, which is to say plants which reach for their star as they seek the warmth and the light and the energy from their own sun.

Creation is big, I mean really big. It is beyond most people's comprehension. Even the Creator did "reach for the stars", and according to the Genesis creation account, the Creator managed to achieve that goal on day four.

Chapter 19

Retaining the Information

We can harvest all the information we want, but if we cannot retain true information then we have nothing and have wasted our time. Retention of data is critical. It is critical in each point of calibration done in the creation account, and it is extremely critical in the DNA of plants. Can you imagine what would happen if the information in DNA was not retained on a consistent basis? If you have ever seen a plant that grew in some environment that was toxic to the plant you will likely have witnessed firsthand what can happen when the information is not properly retained or is corrupted in some manner. Mutant plants are unpleasing to the eye and are not safe to consume when their information gets messed up through improper DNA retention.

The same is true for the information that *we* retain. When we foul our information with bad information or lies and inaccuracies of pop-science or erred religious points of view, we then produce mental fruit that is not healthy for others to consume or for ourselves to dwell on.

Storing Up Information and Building the Instructions

As we collect information we store that information for use at a later date. But what will we use that information for? The information we collect becomes a set of instructions for us to follow as we progress in life. If we retain bad information, then that bad information is added to our instructions, and so our next move will be made based upon our previous error-filled instruction set. This is why so many people on both sides of the godless-evolution versus six-twenty-four-hour-day debate get so far off track and cannot correct themselves.

Their mental DNA instructions, which is the information that they mentally retain, has built them an instruction set that has been altered in the genetics of their thinking. They cannot escape those altered mental genetics unless they can eventually overcome the corrupted information and filter it out like the plants do.

Lucky for us, plants tend to self-correct and are generally able to overcome bad information that they might receive as they build their DNA instruction sets. The bad DNA gets filtered in one of two ways. The first is that if the fresh information stored in their DNA is fouled, then it is common that the plant will not thrive and will likely die and not reproduce, thus saving its kind the disgrace of massive defects. The other is that the DNA in plants has an ability to reject most of the bad information it receives as it builds the next generation of DNA and stores it in its seeds. This is another testament to "the fruit tree yielding fruit after its kind which may have seed in itself upon the earth"

Thinking Outside the Box to Think Inside the Box

Our "mental DNA" is not some sort of genetic cellular DNA that we have no control over. Our *mental* DNA is the instruction sets that we *allow* into our thinking. You cannot understand something when you reject understanding. It might sound odd, but many of us do reject understanding when we choose to fill our mental DNA instruction with inaccuracies and lies that do not match with what we witness every day with our own eyes.

We have been given an overwhelming set of data, and that same data set is given to us every single day that we step outside or water our in-house plants. But our inability to think outside of our little *mental* DNA instruction boxes causes the contamination that we have already accepted into our mental DNA to be trapped inside of our mental DNA boxes—along with us.

For us to clean and purify our mental DNA we must think outside of the box to think inside the box. When you look around the world, stop to think that—what you think—could be wrong. Sure, it might be correct, but if you fail to think outside your own mental DNA box, then your box becomes smaller and smaller and tighter and tighter. And eventually it will choke you out and destroy your *mental* DNA.

When we function like the plants and think outside the box we can spread our healthy mental DNA to others, for them to process and add to their own mental DNA.

Remembering the Evidence

What you remember is your mental DNA. You have been given vast amounts of evidence. Remembering that evidence is important, but it is not as important as remembering it **properly**. This is why thinking outside of the box is so important. People often take this as—they have to somehow *defy everything*—but that is not at all the case.

Thinking outside of the box is merely a way of verifying your information by testing it against other people's contradicting thoughts. But when we choose to remember the evidence in an improper way, then we nail our box tightly shut. If plants did this, then there would likely be very few variations, if any, of each "kind". While plants do not have souls as we do, we are very much like plants in these ways, but it is our chosen instruction set that guides our life **and** guides us in what questions we will ask.

Following the Instructions

Many people balk at "following instructions", especially when it comes to the Bible, yet those instructions are there to protect us. We have been following most of those instructions for thousands of years without realizing it in our "western culture".

We often have bad mental DNA instructions due to the sources of information we choose to allow in us, but often for no good reason we outright reject good and true information that we come across. In our defiance, we cheat ourselves out of being able to think outside of our own box. In doing so we often imagine that we are thinking outside of our box, but we are not. We are thinking outside of other people's boxes who have good information that we might need.

When we deliberately defy good instruction, we are then technically following bad or fouled instructions, and are failing to grow and learn. If we allow bad instructions into us then we will inevitably follow paths of false information in the future.

What would the world be like if plants did not adhere to the calibrations set forth on day three? What if they could choose to follow bad instructions like we do? It would likely result in a similar effect as with us humans. The plants would possibly mutate to a fouled state, ultimately becoming unusable. They might never have developed the wonderful diversity that we see today due to their abiding by the calibration of "after its kind". The kinds would have likely died out and no plants would be here today–and neither would we.

The ability to follow good instruction is perhaps our most cherished human ability because it gives us hopes and dreams and it also offers us the ability to research the heavens and earth, as well as plant life. The instruction we choose to follow dictates our individual and our collective scientific progress for all of humanity.

Chapter 20

The Sprout

Information sprouts from our minds like a plant sprouts from the ground. Your mental DNA is planted in your mind and it will grow and become what you guide it to become. *You* alone are the master of your mind. But just as the plants were given base instructions in "after their kind", so, too are we humans given base instructions. Those base instructions are our basic human logic and our ability to seek and know Truth. Our mental DNA base instruction is inherited from the Creator, and when we foul that base instruction we can get far off track as is witnessed in pop-science and six-twenty-four-hour-day creation.

The First of Its Kind

How do we determine "the first of its kind" in any evolutionary sense? For instance, according to the Bible, the first of our kind for us humans is the Creator, as it is the Creator that we are from and in whose image were made. But with plants it's different. The plants are not said to have been created in the image of the Creator, instead the earth was to "bring forth". So, with plants, was there a "first of its kind"? "First of its kind" is not a biblical statement, but is an underlying curiosity that many people have regarding creation.

It is unlikely that one plant was a "first". A single kind or variation could have occurred first, but that kind or variation likely occurred in many places simultaneously all over Earth. We see this also in the fossil record. Thus the "first" was not a specific flower growing in one spot that slowly spread its seed around the globe. Rather, with plants, the "first" would have simultaneously begun in many places based upon the day-three creation text along with our observations of the geological record. However, the geological record is generally irrelevant in this case.

While the geological record does tell us a great deal about plants, what we do not know about that record is the time scale, and further, if there is additional recording of events deeper than we have been able to look in the layers. If we can believe that all the layers we see were laid down over time and were not instantly created by God as we see them today, we also have to wonder if there are more layers there than meet the eye.

Plant Messages

Do plants have souls? Can they think? It is unlikely that plants have souls or are capable of thinking. But some years back a researcher used a lie detector on plants and he discovered that plants react to human intent. He discovered this accidentally and was *not* testing to see if the plants lie, but, for some reason, he attached the lie detector's sensors to a nearby plant and noticed, by chance of events, that when he did certain actions, the plant would move the indicator needle on the lie detector.

Now this experiment proves little if anything, and I am not aware of it ever being repeated in a lab by others due to the nature of the publicity someone might get in trying to publish such information (Think little mental boxes here, and being mocked for thinking outside of the box).

Nonetheless, the operator of the equipment did notice this and carried out subsequent tests that all indicated similar reaction by the plants. I do not discuss this specific plant issue in the book *Understanding Prayer – Why Our Prayers Don't Work - The Prayer How-To Manual*, but I do touch on the repeatable laboratory experiments and the brain scans that we do with people where these scans clearly indicate that our bodies emit and receive electromagnetic signals that are closely associated with our brain activity.

The point I am making here is that if plants are somehow physically sensitive to electromagnetic signals produced from our brains, then our thinking or thoughts could have some effect on the plants or possibly even them on us. Not in a soul-to-soul manner but in a logic-state of *plus* versus *minus* or *good* versus *bad* or a *harm* versus *nurture* manner.

Plants do hold messages for us in their DNA instructions, but do they have any other continuity to us humans in the bigger picture of creation?

Plant Husbandry

In a point later in Genesis than we have been discussing "God planted a garden" and placed "man" there to "keep it". Plants were intended to be cared for by mankind. So, while the plants feed "man", they also are to be cared for by "man". We have natural habitats all over the world. These natural habitats are often lush and green and filled with a diversity of beauty and color; but there are also habitats that are barren and mostly void of life. Yet when we humans choose to, we can take these barren habitats and care for them and cause them to flourish, overflowing them with plant life. This has been occurring for thousands of years as we humans have decided to care for any part of a barren garden and make it flourish.

This is another testament to the synchronicity that true science and the creation account share. There are many such cases that if we choose to look we will see that the world that we experience today is undeniably in full agreement with what it says in the Bible's Genesis One creation account.

Chapter 21

What Came First?

The chicken-versus-the-egg issue is truly an interesting topic, but it often leads to bickering between six-twenty-four-hour-day creationists and godless-evolutionists. This *cause*-and-*effect* issue is the root of godless-evolution. The primary reason that many of us have abandon the logic laid down in the Bible's creation account, is that the little mental boxes of many preachers have failed the people on a scientific level and continue to do so to this day via sermons and any perverted post-reformation Bible versions.

When people alter the intention laid down in the authoritative versions of the Creation text in Genesis by changing words within it, they inadvertently force others who want to be more discerning to abandon the illogical interpretations of the text, thus causing them to subsequently abandon the Bible. When we analyze and then adhere to erred interpretations we then understand the Bible as if it is wrong and does not agree with what we see in scientific evidence.

When preachers insist on ideas, such as seeds came first and were planted, or that the waters that were separated on day two were the clouds and the ocean, it makes no scientific or logical sense whatsoever and thus invalidates the remaining text. And if Genesis One is invalidated then so is the remainder of the Bible in the eyes of many readers.

"What came first" not only applies to the-chicken-versus-the-egg and the-plant-versus-the-seed, but it also applies to the order of creation. The *GOD'S WORD* translation of verse six says "So God made the horizon and separated the water above and below the horizon. And so it was." which is very different from most other versions and from the Douay version which says, "and god made a firmament and divided the waters that were under the firmament from those that were above the firmament and it was so". So as you can see, these sorts of changes in the order of events cause confusion because it creates an incorrect mental picture of Earth and oceans and clouds. Our erred mental picture occurs even if that was not the intended original interpretation; however, it was the intention and understanding of the translator of the *GOD'S WORD* version. These sorts of translations can be considered junk translations.

They Came First

These sometimes-little changes in the Genesis One text obscures our view of what the text actually intended to convey. We must use logic, human experience, and science or discovery to begin to actually properly understand the Genesis creation text.

While it is very obvious that the order of the creation events is key in it being a credible account, so is the order of the-plant-versus-the-seed. If anyone supposes that God made seeds spontaneously appear as we know seeds today, then they are likely fooling themselves into believing lies, or they are thinking inaccurately. Everything made by humans or the Creator occurs through a process, and no one has ever witnessed otherwise, nor is it otherwise stated in the Bible.

To imagine that the Creator just spontaneously made seeds appear is like imagining that the engineer just suddenly spontaneously made the blueprints... POOF! Blueprints! No, the engineer has to plan and think and imagine, and then once that part is mostly complete, he can begin to draw out or write out those plans or instructions in a concrete manner where the packet of information (the blueprint-DNA) can then be followed by the builder. Since we are created in the image of the Creator, we must realize that the Creator has to start somewhere through, in some way, laying down an instruction set by utilizing the now already gathered materials.

This process could have taken a great deal of time. This is like the engineer finally completing his work and then handing the blueprints (or the subject's DNA) to the builder and saying, "Build this" or "bring forth".

Their Purpose

Once the order has been given, it is then that a building's instructions will be followed and all of the initial workings will occur underground as the foundation is slowly assembled and is ready to support the structure that will be placed on top of it. At that point, the foundation or roots of the building have tapped into the water system, into the electrical grid, into the septic system, and communications system, and into a materials supply system, all of which are similar to nutrients for the building and will all allow the building to progress, grow, and thrive.

And in general, as long as the sun shines, the structure will continue to grow above ground as the builder assembles each bit of material. When the building reaches maturity it will be decorated with colors and other useful items, all of which will bring a bit of joy to the future occupants.

Seeds have the purpose of carrying the DNA blueprints on to each subsequent generation of plant. And the plant has the purpose of nourishing, protecting, and beautifying. As to their original purpose, we cannot specifically know for sure, but if a Creator does in fact exist and is capable so as to carry out the events in Genesis One, then we have to assume that plans were being planned in advance and subsequently carried out. While this is the likely course, we must also realize that some of the creation events were likely done in a somewhat prototyping manner, where one event was carried out, and it is the analysis of that event's result that directs the next event or next action taken.

Having done a great deal of prototype work in designing and building various machines, I can see no other way a Creator would approach creating that which never before existed. This is especially true when we consider that there was nothing beforehand to compare to. It was likely all a *first* effort every step of the way, with no past samples to evaluate other than the completion of each previous event's work result.

Everything has purpose. Instructions have purpose, and the resulting items have purpose, but does that "purpose" have any *meaning?*

What Plant's Mean to Other Life

Here's another chicken-versus-egg or plant-versus-seed issue: Not to get ahead of a following volume of *The Science of God Volume 3 - Day Five and Day Six - The Creatures - Revolution or Evolution*, but animals are typically not carnivorous, meaning that most don't eat meat and thus their only other food source is plants. The plants came first and fed the animals, but did the Creator make the plants for the animals to use, or where the animals designed to eat plants because that is all they could eat to acquire their needed energy and nutrition?

Either way, no matter how we want to view things, we and the animals and sea creatures will all die without plants, and that is another clue as to the accuracy of order listed in Genesis one. But plants are far more than just food. Plants are protection as animals blend in and hide in them. Plants are a dwelling place and supply materials for the animals' homes for comfort and for protection from weather. Plants feed the animals. Plants even feed the plants. Plants also offer diversity of colors for animals to enjoy. But what do plants mean to us humans?

What Plants Mean to You

As with animals, plants mean a great deal to us humans, but we humans seem to have a different or unique ability to appreciate them. To you, plants mean a home, typically built of wood from tree plants. They mean vegetable plants for foods on your table. They mean fiber plants for the clothes on your back. And as we understand it, plants mean oil for fuels and making materials like fabric fibers. Plants are our livelihood as we make a living caring for them, harvesting them, modifying them for wood, for food, for decoration, for fragrance, and so much more. Setting aside God, plants are everything to us because without them we would promptly meet our end.

Imagine for a moment if somehow animals were created, whole, as adults in mass quantities. I realize that this stretches the imagination, but now I want you to imagine these adult animals without any plants, thus making them all carnivores having to eat other creatures to propagate. Using logic, how would that work out? Now imagine us humans coming onto the scene having to kill animals for our food because there are no plants for us, because plants do not exist in this mental exercise.

If animals and humans could only eat each other, because no plants existed, eventually we would diminish as the energy in us was utilized through our movement and growth. The quantity of animals and humans would slowly diminish with each generation until only one was left, and when the last one was standing alone there would be nothing left to eat and thus they would eventually die. Since plants would not exist, the excrement from all creatures would not produce any food.

Not only do plants mean beauty, protection, fragrance, food, and joy to us, but they also mean life and increase. The plants harvest energy from our Sun and they take that energy and turn it into nutrition that allows us to live and breathe and grow. We humans can then multiply ourselves through having children and we can repeat this and increase our numbers only because of plants. And *that* is what plants mean to you, and it is a truly unique and ingenious system!

Chapter 22

We Must Ask Why

Why is everything here? Did it just happen like big bang proposes? Or is something bigger at play here? Our ability to wonder is the most magnificent human feature. Did *you* ever wonder?

Did You Ever Wonder?

Asking if you have ever wondered about something is a nonsensical question because any question that ever arose in our thinking is "*wondering*". But things change a bit if I ask, "Did you ever wonder why?"

Asking *why* implies reason, asking *how* does not. Godless-evolution, and big bang as well, only attempt to address *how* because *why* is not allowed in that line of thinking and rationale. And yet, we all do wonder *why* about things. Sometimes we are really seeking a *how* answer, but mistakenly state our *how* question as a *why* question.

Separate the *why* from the *how* questions as you consider the creation topic, because nowhere is understanding that there is a difference more important than in the creation topic.

Think about the notion of plant evolution; if plants did spontaneously begin through some accidental chain of events, then can we ask *why* it happened? Think about it. I ask you *why* did plants begin and you go on to scientifically describe a theory as to what all occurred in order to give us the plants that are here today, but that is not *why*, that is **how**. *Why* versus *how* is a fine but cloudy line in our heads. *Why* is unique. *What, where, when, why*, and *how*. *How* is the, we can call it, "scientific" description of the way the *what, where*, and *when* are carried out for the *why*.

The *why* has intent and it technically cannot occur without discerning *beings* being involved. Did you ever wonder *why* you ask questions at all?

Seeking Answers

Our ability to ask questions is directly related to our desire for answers. For thousands of years we humans have been recording our quest for answers. We have found many *how* answers, but we struggle finding our *why* answers. We struggle even more when we try to answer a *why* question with a *how* answer. This is because *how* answers can never properly satisfy *why* questions.

Seeking answers is something that we all do. But separating our *hows* and *whys* is something that few people realize they need to do. How often have we ourselves thought of or heard someone say "*how* could someone do that?" *How*? That's simple, a criminal simply walks into a place and carries out the act of doing their crime using the tools or weapons that they chose. *Why* they would do that, on the other hand, requires a quite different answer and probably contains a lot more information about the person and their upbringing, life, troubles, and attitude.

As you can see, *why* is human or requires some form of intelligence, on the other hand, *how* is mechanical or procedural. Knowing and realizing that there is a difference between *how* and *why* will bring you to a place of being able to more quickly extract the information you desire to know from the answer you received.

Asking the Right Questions Versus Asking the Questions Right

When we ask questions, it's common to switch out the *hows* with the *whys* as mentioned in the last section. But a part of our problem is, even if we have our *hows* and *whys* properly laid out as we present our questions, we still often receive a *how* answer for a *why* question and vice-versa. So, while we can ask the question correctly by properly forming our thoughts and having the *whys* and *hows* clear in our own mind, the answer can still be a mismatch for the *how* or *why* because of the way the person offering an answer understood the question or the entire topic and their understanding of life and science etc.

You can *ask the right questions* all day long, but until you *ask the question right*, you will likely get incomplete or altogether incompatible answers. *Asking the questions right* has a lot to do with where you intend on getting your answers. If you're asking people questions, then to *ask the questions right* involves you doing a quick analysis of the person from whom you seek information, and you must realize that the way in which you ask a question can have a great impact on the answer that they return to you, even if they have their *hows* and *whys* in order.

So, if you ask someone a question about, oh let's say, finances and they don't understand the financial lingo, then you might be better served to turn that question into a sort of sports analogy if that is their thing. This will allow them to better relate to the information that you seek, and you are then able to translate their answer into your needed terminology. This is an easy enough concept to grasp, but what about when you must question yourself or the Earth and its geological record? Then what? Only you can really answer that. But there is one thing you can be sure of, and it is that when you are seeking a red rock, then you will find many in the red color range, and all other rocks sort of vanish from your view and become plain old rocks to you. So, if you're looking for a reddish rock, then you need to have that clear in your mind, otherwise you are basically looking for rocks in general.

There can be a difference such as in wondering *if* red rocks exist, versus *where* might red rocks exist. One asks *if* there are any red rocks, where the other assumes that the red rocks exist but wonders **where** they might be found. The difference in results is usually quite profound.

Seeking the Light

The interesting thing about asking questions is that even when a question is formed poorly, we can still usually obtain the needed information, but that's regarding simple things. More complex or hidden types of topics require more refined questions, and once we have perfected our question, then the answers seem to come more quickly.

It is our recognition of that which we are asking about that allows us to see the answer. You could walk past a reddish rock ten-thousand times and never even notice it existed. In fact, it is likely that if you tripped on the reddish rock and stubbed your toe you would likely utter some language against "that #%$! rock!" not even caring or noticing its color. But let's imagine that you went to study geology and found that there is a certain rare kind of rock that is reddish. Then you would suddenly either recall the rock that you tripped on or you would recognized it as something unique the next time you passed by it.

The proper formation of questions is paramount in our ability to receive the answers we seek. This is true of issues surrounding

God and salvation for sure, but it very much is a part of the Creation topic of Genesis One, of science, of geology, and of botany, and it includes life in general.

Many Christians desperately want to believe the Bible to be a perfect God-given document, but they cannot reconcile the language and terms we read in the Bibles that sit upon our end-tables. Genesis, for too many of us, simply does not reconcile with the science we packed in to our tiny mental boxes. We seek to understand, but we cannot recognize the truths that we walk past every day that scream out to us for our attention to prove that which we seek to know and typically expect, but all too often biasedly ignore.

When you can form the question properly because you have the right seed of information, then things change a bit and the fountain of knowledge begins to flow for you. This goes back to the point made earlier that information we allow in our mental DNA will determine our next move, or in this case our next question. These choices matter.

Chapter 23

Our Choices and Plant Choices

Plants have little choice in what they will do. Plants were given an instruction set when they were commanded to be brought forth by the "Earth". And the plants do not veer from that DNA instruction unless acted upon by nefarious external forces. But us humans have choices, lots of choices! While our cellular DNA will dictate our form and functions, our thoughts and actions on the other hand are unique from the plants, because *we* can choose.

Things like answering our *whys* or our *freewill* are tough to grapple with when looking at things scientifically. Science does not, and cannot, answer these *why* questions. Although, with our ability to mimic intelligence with computers, we imagine that we can partially answer these questions. But there again, we are only closing in on the *how* and are still not addressing the *why*.

As we traverse our journey on our quests for answers, we speculate using computers and modeling and calculations to "prove" our theories.

Anything for a Buck

As we choose the paths of our human quests, we often end up doing so for money. We get ideas in our heads and refuse to let those ideas go even when we see some error in them. We get so stuck on our own theory that we refuse to entertain other theories and test them against our own to see if they are a better fit to what we see in nature. Often these ideas or theories are presented through our books or papers that we submit, or maybe in an interview we do. None of which is problematic until two things occur: The first is our **un**willingness to release our errors, and the second is our greed to advance our theory for financial gain regardless of how flawed it is. For far too many science pontificators it's all about the money.

Imagine if seeds refused to follow their growth instructions in their DNA to bear fruit, this would cause a great deal of hunger. The same is true of our refusal to bear good fruit with our science. It keeps people hungry and wanting more, but after they get tired of trying to find fruit on the weeds that they have been offered, they go elsewhere to forage for mental nourishment. Science is certainly guilty of this particular anything-for-a-buck sin, but so is religion.

Anything for God

We see preachers requesting money from their viewers as they preach "messages from God" using the Bible to make their case. Often what they say makes sense and is good, thus helping people on a mental and/or spiritual basis. Many preachers would do anything to serve God and so would many of their viewers who are dedicated to God. This is all good and is as it should be. Until...

Yet, what are we when we are "serving God" and will do anything for God, but are preaching incorrect information to the viewers?

Is it okay for a preacher to preach six-twenty-four-hour-day-creation if it is **not** accurate? Is it right that they pick a side and disregard science and all that we find in nature for an inaccurate interpretation of Genesis One? Many who claim that they would do anything for God are sometimes liars. Harsh? Yes. True? Yes. Of course, this does not apply to all, but rather only *some* people and preachers. If you claim you will do anything for God, then your first course of action should be, above all else, to seek Truth from God and to be *accurate*.

But if the "truth" you might think you may have read in a bad translation of Genesis makes you have any illusion that plant seeds came first, or that the Sun was made on day one, or that the first separation of "waters" is the clouds and the oceans, then you do not have Truth in you and you are believing lies. If this is your case, you can be assured that when your children try to explain that point of view to teachers or friends, or if they attempt to explain it on any form of media, they will promptly be torn to shreds by a very vicious audience and will be humiliated for their errors and lack of understanding.

If you're the type person who will do anything for God, then make sure that finding Truth is at the top of that list. And while the Genesis creation account might seem unimportant to you or to others, it is the very foundation of our existence and it is directly attacked when the errors in some Bibles allow the Creation account to be discredited in the minds of many people through inaccurate translation. We humans will sorely regret this error if gone unchecked, as it has already undermined the beliefs and understanding of far too many people.

Anything for the Agenda

In our contemporary world, we hear a lot about agenda-driven promotion of ideas. And these agenda are typically presented as being negative or somehow unacceptable. But everyone has some sort of "agenda". Anything that we want to promote or discuss is a part of our agenda. The question isn't whether or not we have an agenda or if the agenda is *perceived* as bad or good. The only real question is whether or not the agenda is based upon actual *Truth* and *goodness*.

An agenda can be based upon the Bible and is therefore based upon good, but if that agenda lacks the truth that the Bible is based upon then it is *not* a truly good agenda. What something is based upon does have bearing on the agenda or theory, but it does *not* dictate the good or bad or the Truth of the agenda.

There are many pop-scientists in various fields surrounding the origins issues, but most of them have an agenda to discredit the Bible. If your agenda is to destroy and discredit your competition in order for you to be viewed as victorious, then you might want to reconsider your own theories, because all theory based upon Truth will eventually prevail.

What is Provable?

Provability is often in the mind of the prover. If we seek to destroy our opposition, then it is likely that our own theories lack any provable evidence. For instance, godless-evolutionists often claim that no Biblical thing can be proven, and go on to claim that scientists are not afforded this luxury because they know that their data will be scrutinized by hosts of their peers. But this is simply not true.

Sure, the data might be scrutinized by their peers, but do their peers agree with their general premise? Are their peers on the same quest? Isn't that like two six-twenty-four-hour-day creationist having a conversation and telling each other why six-twenty-four-hour-day-creation works? Yes, it is!

Frauds will not be rooted out and public humiliation cannot occur when an agenda is shared by many. And many people sharing a single agenda does *not* make that agenda correct. Attacks on the opposition that claim a peer-review type argument will often use strong language and derogatory and slanderous comments in their unscientific attacks. Taking this easy route of attack and slander is an arrogant approach by people who cannot prove their data with true science.

There is a difference between referring to someone as ignorant, and referring to them as an ignorant imbecile. Ignorant denotes that they are ignoring information which is all too often done by evolutionists who further go on to claim that their information is "undeniable fact" and then state that their information will be willingly changed when other better information comes along. This is simply not true, and further it points out that their original point of "fact" that they have now changed was, in fact, **not** "fact".

I once heard a pop-scientist commenting on how scientists of the past tried just about anything to save their beloved theory of "aether". This makes sense, because if a theory partially explains things and we have a lot of research time invested, then we want to hang on to our investment. But the interesting thing about this particular situation was the conviction that the person had, and how amazed this particular person was that the past scientists simply could not accept that aether did not exist. What is interesting is that the very same person cannot accept that the universe likely did not bang itself into existence and evolution is not Godless, but was and is, highly likely, intelligibly guided.

Our willingness to alter our perception of the apparent data is a noble and proper way to conduct research, but calling what we believe today about big bang "fact" and calling people names to hide our apparent ignorance to point out the apparent ignorance of others does not make what we claim as "fact" to be actual fact.

It's easy for anyone to post their opinion in some public manner and get accolades from their peers and from those who want to agree with them, but it is quite another to actually be able to back up the data being used to proselytize their minions.

It doesn't matter if this is a science or religious perspective, the same is true.

If you want a bit of truth, consider this: You must be willing to toss out all of your evidence and information that you have believed to be true and factual, and then test new evidence presented in order to be able to test your own original beliefs.

I cannot recall who said it, but I once heard a godless-evolution proponent say something to the order of, any scientist has to be ready to trash all information that he or she has believed in, and consider it crap, and then move on in the face of new evidence. This was surprising and was a very welcome sentiment, but seldom is this the case in pop-science.

Proving something without full consideration of opposing views is not really proving much at all, unless the proof is very concrete and compelling and is *obviously* true.

Chapter 24

Plants Live On

Truth is eternal and will live on no matter what we do or say or think. Our *beliefs* are irrelevant to Truth. Truth about plants is what actually occurred and our godless-evolution theories or our six-twenty-four-hour-day theories are no challenge for Truth. Our flawed theories are easily thwarted outside of the comfort of our little mental DNA boxes, and the same is so when we are out of the comfort of our minions who will indulge us in our erred theories.

Let us all be thankful that the plants live on in Truth and follow the calibration laid down for them. If not for that, none of us would be here to wonder about anything. What we should all do is to somewhat imitate the plants and be reliable and stable and allow variance as we grow and learn, rather than crawling back into our deceptive and unstable little mental boxes every day–mental boxes that have been contaminated by an onslaught of nefarious information.

How Do They Even Bother?

If not for the plants, our world would look like our Moon or like Mars. Imagine living in that world barren and void of all plant life. Just look around and try to not see plants—They are everywhere! How do they even bother? How do plants continuously keep regenerating? They do so through the DNA instructions that they were brought forth with. Through all of the steps that are discussed in this book, the plants that we know and love today exist. And through our care and experimentation, those same plants will continue to robustly feed all of the world that is allowed to obtain that food.

Plants do this freely at no actual cost to us. If we find a seed we can plant that seed and it will grow and eventually provide food for us—<u>free of charge</u>! But unless we tend to the plant, or pay someone else to tend to the plant for us, the plant might not produce much if any food good enough to eat. Or it might be overcome by insects.

Plants reproduce their DNA instructions and produce new replicas of themselves all without our help and they do this through their DNA instructions given when they were initially brought forth. But is there a *why* in them at all?

Why Do They Even Bother?

Can you imagine if plants could *choose* to live on or choose to produce food for *our* sake? The plants would take one look at all of our wars and infighting and shrivel up and die to let us to fend for ourselves.

Why do plants even bother? Plants bother because plants do not have a choice. They have obvious design and function and they follow exactly the guidelines by which they were to abide as instructed in Genesis One verses eleven and twelve when they were brought forth.

But in all of this, we look at plants and still wonder why things are as they are. We can imagine that it all just randomly happened by chance, but I would bet everything on the fact that if we could travel to other stars and explore, that we would find many instances of nearly identical patterns in the plants with the three major kinds being herb, general vegetation, and fruit plants.

And if we do ever find any near identical plant patterns alive on a distant planet circling a distant star, then would we finally be able to admit to the accuracy of the creation account? What amount of evidence is required for us to take the creation account as the serious but broad scientific account that it is?

There is one prediction that I will make here and it is that if we humans are ever able to travel to other solar systems and find planets resembling the conditions of our Earth, we will find things to be just as they are here. This will be to a point where if you were set down on such a planet, you would suppose that you are right here on our Earth, but you would not be. And as a part of this same prediction we would also likely find that many, if not most, other solar systems would be the same having one or even several planets that support plant life similar to what we experience here on our Earth. We would not only find the three major kinds, but would likely find many similar variations of those kinds that we would recognize as the same fruit here on our Earth such as apples or bananas etc.

Continuing Life

Plant life will continue, and it will follow the DNA instructions that build those plants' patterns and functions. There is nothing we humans can do to stop this even if we try. While we can destroy a single plant, we will never be able to stop them all. Plants will continue, and in doing so they will continue to increase the fertility of earth and increase life on Earth. There is nothing in the fossil record that suggests anything other than plants being incredibly dependable, resilient, diverse, and consistent.

Plants will continue despite our human nature. They will do their plant life and reproduce and they will allow us to mold them into other variations of bigger or better or more succulent food from which we can obtain the nutrition of life.

The Plant Promise

So long as the plants continue to follow their instructions, they will continue to feed us either directly through us eating the plant or their seeds or fruits, or we will eat the plants when stored in the form of meat from animals or their byproducts. Plants will surpass the existence of us humans; there is little doubt about that.

Do plants make a promise to us? In a way yes, but that promise is more of a commitment that is inherent in all of creation. The created pre-days-heaven-and-earth in the first sentence of Genesis One are obviously the substance of everything that followed. And each step, each day, was a sort of method or a way of commitment that was conceived by the Creator. Each of the first four days were broad, but precise, calibrations that are commitments or promises holding it all firm and it does not vary—and we have no evidence to suggest otherwise.

All of the events of the first four days matter to us, but it is when the living things were brought forth that it began to really matter to us humans, even though we did not yet exist. The commitment made by the Creator through the bringing forth of the plants and, more importantly, the command for the plants to produce seed after their own kind, is our food, it is our sustenance, it is our physical life.

Thank you plants and thank you Creator for the everlasting and diverse beauty we find in these plants and for the variety of food which they produce for all of us to enjoy!

***Oh Mighty Creator, Please Bless the Plants
and teach us to care for and nurture them
while they care for and nurture us!***

Announcements

Rock the Boat with Layers of Truth

Do you believe that the entire world flooded roughly four thousand years ago and that a man named Noah built a large boat to save a small remnant of human and animal life that would repopulate the entire Earth? This is the belief of many Christians, Jews, and Muslims, but then we have those who believe that the entire story was written thousands of years ago for entertainment only.

Could either case be true? Is either realistic? After all there is a lot of evidence of catastrophic worldwide flooding. But then there are those making the point that there's not enough water on Earth to cover the mountains. So, which, if either, is it? If either case were proven to be undeniably true it would have major impact on opposing perspectives. If it never occurred, it would devastate most Bible-based religions. But how would it affect modern sciences if it was proven true? It would force every scientist to face a reality for which they have not been educated.

Take a journey through these and other Biblical flood questions and consider the perspectives presented in *The Science Of God Volume 5 – Boats, Floods, and Noah – The Deluge*, a truly logical scientific explanation of the viability regarding the Biblical flood of Noah's time.

Search: The Science Of God Book Volume 5
SayItBooks.com

Announcements

Rocking the Cradle of Life
A Decent Account of Descent

Have you ever wondered if humans actually did evolve from apes? Or maybe, if we were specifically created, then how might have that occurred? There sure are a lot of opinions on the evolution versus creation topic. And too often these views use confusing technical jargon that few people care to learn or have ever even heard.

The answers to the questions you might have are, in many cases, the same answers that many other people seek. When you have solid answers that are difficult for someone to thwart, it's good to share those answers so that others can also feel confident with their own understanding of the arrival of mankind and the level of importance that it has in their own lives.

The Science Of God Volume 4 - Evolution versus Man – In Our Image takes a deep but simple dive into the human evolution versus human creation debate using simple language that everyone can understand and enjoy!

If you have thoughts that you have been reluctant to share, then suspend your thoughts for a bit and open your mind to consider the perspectives and evidence presented in *The Science Of God Volume 4 - Evolution versus Man – In Our Image*. You will acquire a much clearer view of the subject as you read the various points made in this engaging book about the arrival of mankind.

Search: The Science Of God Book Volume 4
SayItBooks.com

Announcements

A Fishy Flying Account Crawling Out of Nowhere

Have you been trying to share your views in the evolution versus creation debate but are thwarted at your every utterance? Are you reluctant to speak up and share your opinions because you're not sure what is or is not true? Sometimes we might even wonder if we should even bother pondering these things at all since no one can ever truly prove their theory to a point of it being "undeniable fact".

Take heart because there are more possibilities than are offered by most people on either side of the discussion. Bystanders often observe the views from both sides of the debate and will then consider those perspectives and try to balance them using logic, but we often fail to achieve that logical balance.

Balance is achieved by many people, but it is typically compromised in order to arrive at an agreeable viewpoint. Ignoring facts in this way is no way to discover truth.

The Science Of God Volume 3 – The Creatures – Revolution or Evolution will not force you to ignore any true facts, and will guide you on your quest to see the clear path to how creatures came to be. God or Evolution? You decide, because everyone is welcome in the discussion!

**Search: The Science Of God Book Volume 3
SayItBooks.com**

Announcements

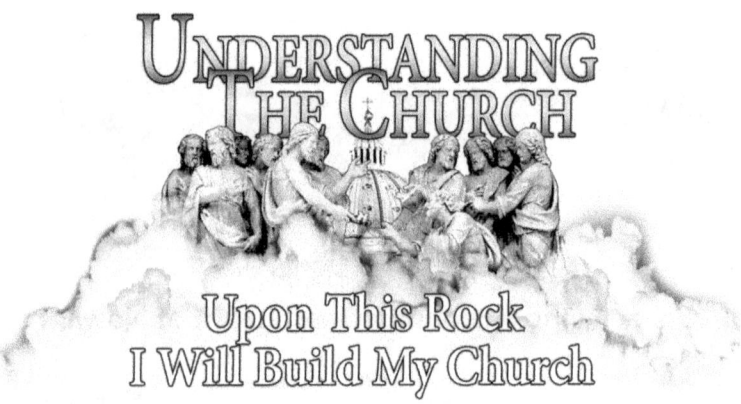

Church in the Lurch - a House Built Upon Sand

The Church is rapidly dying, and much of the clergy in recent times have been doing it more harm than good. People are fleeing from the Churches as they seek a religious perspective that fits a modern worldview. Should we revive this old Church and try to save it from its own demise? What exactly is "The Church", and who or which of the many religions is the official caretaker of it?

The Christian religions of the world have done their fair share of damage to themselves and to the world, but in the bigger picture, they have done more good than damage. Saving the Church is probably worth our collective efforts because the Churches are perhaps the most charitable group of organizations that existed throughout history and even up to today.

The main reason that the Churches are in the rough condition that they are today is due to a lack of understanding by clergy and congregation. We can overcome this dark era of the Church and revive it only through *Understanding The Church*.

Understanding The Church will help you in Bible study, or even to simply better understand the Church. But most importantly, *Understanding The Church – Upon This Rock I Will Build My Church* will help to revive this dying patient.

Search: Understanding The Church Book
SayItBooks.com

Announcements

The Cornerstone of Moral Civilization

Was Jesus really the "Savior"? Did Noah really save humanity from extinction? Did Adam and Eve really get evicted from the Garden of Eden? And what does the word "Bible" mean anyway? When studying or even just reading the Bible, many questions arise to a point where the Bible can be confusing. But when you have certain information before you begin reading, it can instantly propel you to a deeper level of understanding by nothing more than knowing a few key points.

It takes people years to realize some of this information, yet it's not some big secret that only scholars and theologians know. No, this information is for everyone and it's easy to grasp these pieces of information about the Bible and some of the events described within it. Be prepared to have your current views challenged because many things are not as we have been taught.

To truly Understand the Bible, we must open our minds and toss aside all of our biases. Knowing and grasping the often-unrealized basic information presented in *Understanding The Bible - The Bible How-To Manual and The Things We Don't See* brings the Bible to life in a way that shows you, personally, its undeniable relevance to the world, to our culture, and to your very own life!

**Search: Understanding The Bible Book
SayItBooks.com**

Announcements

The Prayer How-To Manual
Understanding Prayer
Why Our Prayers Don't Work

Learn the Real Secret of Prayer

There's a secret that many have tried to understand but failed to accomplish. We pray day after day after day with little or no positive results, causing us to lose faith.

Some people believe that there's a secret method that must be followed to get your prayers answered and receive the things you want in life, but their success is limited, if it comes at all; while others believe that they're not worthy to have their prayers answered. Few people know the True secret, and when they tell us we often misunderstand them.

Understanding Prayer explains, in easy to grasp language, the mysteries behind many causes of prayer failure. True success in your prayers is not measured by how often you pray, how long you pray, or even how badly you want something and how hard you for pray it. True success in your prayer life is measured by *results*!

Understanding Prayer offers you the opportunity to get those results as it reveals the mysteries of a full and robust prayerful connection allowing you solid and repeatable results nearly on command. A little time to read and pray is all it takes to quickly put these sound, true, simple principles to work for you and your family. Gain the understanding of prayer and of how to receive the blessings of financial and mental wealth that can benefit you and keep you free from strife and trouble for years to come!

Search: Understanding Prayer Book
SayItBooks.com

Announcements

THE FAMILY MANUAL
HOW TO BUILD A
STRONG FAMILY
A FOUNDATION OF ROCK

Building A Strong Family Is Easy When You Know!

The world has us believing that, somehow, we are now different than people were as little as fifty years ago. With all of the emphasis on modern behavioral disorders and the mass misdiagnosis of pop-culture diseases, parents have few places to go for information that is true, insightful, and trustworthy.

Strong Family explains, in detail, how family life slowly becomes tainted to a point where our children too often become rebellious and, sometimes, even unmanageable. This even happens to parents who are very loving people.

How to prevent these issues from occurring in the first place is explained in *Strong Family*. But more importantly, *Strong Family* explains the details about how to stop it from progressing further and even how to reverse the damage. *Strong Family* takes a no-nonsense approach to revealing the secrets and mysteries of how parents raise smart, productive, healthy children.

We all deserve joy and love in our family life. Intelligent, healthy, kind children are a right that all parents have, but without understanding the details explained in *Strong Family*, the quality of your children is left to chance and your rights are forfeited. Don't roll the dice with your family. If you want to know the secrets to unlock the mysteries and solutions to a great and joyful family, then *Strong Family* is for you!

Search: Strong Family Book
SayItBooks.com

Announcements

MARRIAGE MANUAL
MAKE YOURS A
Red Hot Marriage
Made In Heaven Filled With Passion and Joy

Learn the Secrets to a Successful Marriage

Have you been trying unsuccessfully for years to tell your spouse the way you truly feel? Are you suffering in a lackluster marriage? Is your marriage on the rocks? Are you planning on getting married in the future? If you answered yes to any of these questions then *Red Hot Marriage* is for you! This straight-forward book covers these and many other common marriage problems and also reveals the causes and solutions for some problems that are not-so-common.

The information in this powerful book, like a true friend, can be at your side with each step you take in restoring your life and relationship to where you likely imagined them to be.

We all deserve lives filled with joy and passion, but our relationships have been tainted by society and by our upbringing. *Red Hot Marriage* strips away all of the lies that we have been inadvertently taught, and quickly teaches you how to regain control of your marriage so that it can be as robust, fulfilling, and passionate as you expected. The mysteries unveiled in *Red Hot Marriage* can have you in command of your marriage in short order as friends and family watch in amazement while you and your spouse walk the path to a strong, vibrant, healthy *Red Hot Marriage*!

Search: Red Hot Marriage Book
SayItBooks.com

Announcements

Take Back Control of Your Life.

If you feel stuck while life unfairly drags you down, then now is the time to take command of your life and learn how to overcome the source of your troubles.

Those around us are often those who hold us back from living rich and robust lives. Realizing that those around us are often those who hold us back helps us to understand somewhat, but in order to free ourselves from their grasp and break the chains that bind us, we need to know *why* this happens.

Cut to the core of your problems with *Hot Water* as it walks hand-in-hand beside you through each detail of the cause of problems while exposing the dirt that society buries us with. This thought provoking book explains the details and how most troubles come to be so that you can better understand what to do about it, allowing you to take the control of your life away from those around you to place it firmly back into your own hands where it belongs.

Advance to your next place in life and richly and robustly live your life filled with wealth and joy. *Hot Water And Your Perceived Identity* assists in gaining full control of your life to change your future forever!

Search: Hot Water Book
SayItBooks.com

Announcements

Theoretical Physics for Everyone!

Can we go back in time? Can we break the light-speed barrier? Is our Sun turning into a black hole?

Explore the secrets of the Universe with a new approach to science that sheds a greater light on pop-science. Mysteries of the Universe are revealed in this easy to understand insightful, in-depth, and thought provoking book about the science of astrophysics.

Theoretical physics can be more than mere theory when the theory is sound. You don't need to be a rocket scientist to understand most of physics; everyone is welcomed in the quest to discover the mysteries of the Universe!

This breakthrough book exposes errors of modern science in the same way that Copernicus, Galileo, and Newton did centuries ago. Are Einstein and Hubble amongst the group of gifted minds that set forth our understanding of the Universe, like those of centuries past. Or are Einstein's and Hubble's theories wrong? Explore these and other questions in *Bending The Ruler - Time Travel, The Speed of Light, and Gravity* and learn how to become one of the great minds that discover the mysteries of the Universe!

Search: Bending The Ruler Book
SayItBooks.com

Announcements

How to Win When You Think You've Been Beat

Feeling grateful is a bit of a struggle when we face tough times in our own lives, and avoiding depression during those times can be tricky. The world cares little of us when we face our own personal struggles, in fact life kicks us when we're down. You've probably experienced the world caring little of your past or present problems, so looking to "The World" for rest and peace is typically of little help.

It doesn't have to be this way! You can change your disposition, and thus, change your future! It's no big secret and it's not difficult, but "The World" won't tell you that, so very few people ever get to hear or understand this simple "secret".

It's amazing to see the people and situations you can attract into your life when you find your own proper perspective, and once you find it you will not want to let it go! Days that test you to your limit become far easier to overcome, making every future test easier than it otherwise would have been.

Simply understanding a few key basics can change your direction in life in short order and can make life a whole lot more peaceful and Joyful! Let *Thank You God – Finding Gratitude in Hard Times* be one of your keys to peace and Joy!

Search: Thank You God Book
SayItBooks.com

Announcements

When You Dream... DREAM THIN™
The Weightloss Repair Manual

Learn How to Lose Weight While Sleeping

How many people do you know who exercise and still can't seem to lose weight? Has that ever happened to you? As a matter of fact, because we don't know the vital secrets that are shared in *Dream Thin*, many of us actually end up *gaining* weight when we exercise.

Do you hit your weight loss goals? And does your weight stay off when you do actually lose some weight? Even many doctors miss the *real* answers to weight loss. If you doubt this, then simply look at the waistlines of many medical doctors and nurses.

Weight loss is easily mastered when you understand a few basic principles. We often go on fad diets or follow the orders of our doctors, only to put the weight back on even faster than we lost it. Many of us suffer from unnecessary disease, and some of us will die too young.

Dream Thin does more than simply share answers to weight loss mysteries. *Dream Thin* explains the important details of *why* and *how* weight loss connects to mind and body. The information in *Dream Thin* allows you to make weight loss permanent without having to try so hard. Don't make more of the same empty promises to yourself each New Year's Day. Instead, quickly and easily change things today and make all of your tomorrows better with *Dream Thin* while still enjoying all of the foods you eat today—and yes, even fast foods!

Only you can choose if you want spend your hard-earned money on medical bills and funerals, or if you would rather spend your time and money looking great while being out and about and enjoying life with friends and family as intended.

Search: Dream Thin Book
SayItBooks.com

Announcements

Volume 1 - The First Four Days

Is there a God? Did we evolve? Did everything start from a big bang? These questions have been plaguing our minds for many years. Only science-minded people and clergy seem to have the answers. But do they really have any true answers?

Is what we are told by science true? Is what we are told by the Church true? Or are there other better explanations for everything? Did we hitch a ride from Mars, or is that all fantasy science? Was everything created in six twenty-four hour days, or did it all take billions of years to happen? Few people are willing to even fully consider these questions, and even fewer have any coherent answers. *The Science of God Volume 1 – The First Four Days* challenges your current beliefs while asking tough questions of science and of the Church.

For years, Christian after Christian has attempted to argue for God and the Bible's Creation only to fail miserably. Why is this, why is it that Christians cannot seem to win this debate? Often Christians think they are winning the debate only to find themselves at a loss to answer the real questions, and then they get mocked for their poor answers.

Whether you are a scientist or an average Christian and want to discuss the Creation debate, *The Science of God Volume 1 – The First Four Days* is a mandatory read for you. *The Science of God* takes you through the thought process to enable you to speak intelligibly about Creation, the cosmos, evolution, and astrophysics.

**Search: The Science Of God Book Volume 1
SayItBooks.com**

Notes

Notes

Notes

Notes

Notes

www.ingramcontent.com/pod-product-compliance
Lightning Source LLC
Chambersburg PA
CBHW071658090426
42738CB00009B/1570